THE
EXPLOSIVE
CHILD

THE
EXPLOSIVE
CHILD

A New Approach for Understanding
and Parenting Easily Frustrated,
Chronically Inflexible Children

Ross W. Greene, Ph.D.

HARPER

NEW YORK • LONDON • TORONTO • SYDNEY

THE EXPLOSIVE CHILD. Copyright © 1998, 2001, 2005, 2010, 2014 by Ross W. Greene. All rights reserved. Printed in the United States of America. No part of this book may be used or reproduced in any manner whatsoever without written permission except in the case of brief quotations embodied in critical articles and reviews. For information, address HarperCollins Publishers, 195 Broadway, New York, NY 10007.

HarperCollins books may be purchased for educational, business, or sales promotional use. For information, please e-mail the Special Markets Department at SPsales@harpercollins.com.

First Harper paperback published 2005, third edition; 2010, fourth edition; 2014, fifth edition.

Designed by Sunil Manchikanti

All illustrations by Greg Daly

Library of Congress Cataloging-in-Publication Data is available upon request.

ISBN 978-0-06-227045-0

19 20 21   OV/LSC   20 19 18 17 16 15 14 13

In memory of Irving A. Greene

Anyone can become angry, that is easy . . .
but to be angry with the right person, to the right degree,
at the right time, for the right purpose, and in the right way . . .
this is not easy.

—ARISTOTLE

If I am not for myself, who is for me?
If I am only for myself, what am I?
If not now, when?

—HILLEL

Illusions are the truths we live by until we know better.

—NANCY GIBBS

CONTENTS

PREFACE

Welcome to the fifth edition of *The Explosive Child*, which comes about fifteen years after the first edition was published in 1998. What a fascinating fifteen years it's been. The book has been published in dozens of languages. The model described in these pages has been implemented by countless families, schools, inpatient psychiatry units, and residential and juvenile detention facilities throughout the world. It has been validated as a research-based, scientifically proven intervention. And the model has continued to evolve as I receive feedback from those using it and try to delineate its principles and strategies in ways that are as clear and accessible as possible. This edition reflects the most current refinements.

Often people ask, "How do I know if my child is explosive?" There's no blood test, of course. "Explosive" is just a descriptive term for kids who become frustrated far more easily and more often, and communicate their frustration in ways that are far more extreme than "ordinary" kids. To be perfectly honest, I've never been a huge fan of the term. First, explosive implies that the outbursts of these kids are sudden and unpredictable and—this may be a little hard to believe at first—that's not true most of the time. Second, while many behaviorally challenging kids *explode* when they're frustrated (screaming, swearing, hitting, kicking, biting, spitting, and so forth), many others *implode* instead (crying, sulking, pouting, having anxiety attacks, and being blue and withdrawn or cranky and irritable). So, the title of the book notwithstanding, the strategies described herein are applicable to kids who are exploding, imploding, or some combination of the two. The term I'll be using to refer to all of them is "behaviorally challenging" (also not an ideal term, but maybe the best we can do).

What you'll learn in the early chapters of the book is that the terms that have commonly been used to characterize behaviorally challenging kids—terms such as willful, manipulative, attention-seeking, limit-testing, contrary, intransigent, unmotivated—are inaccurate and counterproductive. You'll also read that a lot of the things we've been saying about the parents of these kids—that they're passive, permissive, inconsistent, non-contingent, inept disciplinarians—aren't very accurate or productive either. In addition, you'll learn (you may know this already) that the various psychiatric diagnoses that are commonly applied to behaviorally challenging kids don't provide us with the information we need to accurately understand their difficulties and effectively help them.

This may sound a little strange, but there's never been a better time to be living or working with a behaviorally challenging child. That's because an enormous amount of research on behaviorally challenging kids has accumulated over the past forty to fifty years, so we know a lot more about why they're challenging and how to help them than at any other point in human evolution. The research provides us with new lenses through which to view their difficulties, and those new lenses can help caregivers respond to and help these kids in ways that are more compassionate, productive, and effective. That's the good news. The bad news is that the new lenses can take some getting used to (after all, you may have been wearing different lenses for a very long time) so seeing things through those alternate lenses will require an open mind. Also, the strategies contained in this book can be hard to implement (early on), may be different than the ones you've been using, and may represent a departure from the way you were parented. So you'll need an open mind there, too, along with some patience (with yourself and your child) as you're practicing new ways of interacting and solving problems together.

If you are the parent of a behaviorally challenging child, this book should help you feel more optimistic about and confident in handling your child's difficulties and restore some sanity to your family. If you are the child's grandparent, teacher, neighbor, coach, or therapist, this book should, at the least, help you understand. There is no panacea. But there is certainly cause for hope.

ROSS W. GREENE, PH.D.
PORTLAND, MAINE

ACKNOWLEDGMENTS

In prior editions of this book, I acknowledged my family last in this section. I'm correcting that mistake now. My wife, Melissa, has been keeping life interesting and holding down the fort in our family for a very long time. My kids, Talia and Jacob, keep me laughing and learning and help me practice what I preach. Our dog, Sandy, continues to amaze us with her intuition and unconditional love. They all—along with my mother Cynthia, brother Greg, and sister Jill—keep me focused on the important stuff and have been remarkably supportive through some very difficult times. And the book is dedicated to my father, who died long before the first edition was published.

This book wouldn't have been published without the vision and commitment of my friend and agent, Wendy Lipkind, who died after a brief battle with cancer in 2011. While I still find myself wishing I could talk with her on many occasions, I have a decent sense of what Wendy would have said to me on various topics, so I still hear her sage wisdom whenever I remember to ask for it.

This book is a reflection of the remarkable insight, guidance, book-sense, and mastery of the English language of Samantha Martin. I may never write another book without her help.

My thinking about how to help challenging kids and their adult caretakers get along better has been influenced by many parents, teachers, and supervisors. It was my incredible good fortune to be mentored by Dr. Thomas Ollendick while I was a graduate student in the clinical psychology program at Virginia Tech. Tom has remained a very close friend since that time. Two mental health professionals who supervised me during my training years were particularly influential: Dr. George Clum at Virginia Tech and Dr. Mary Ann McCabe (then at Children's National Medical Center in Washington, D.C.). Lorraine Lougee, a social worker when I was a psychology intern at CNMC, gets credit for pushing me to take a strong stand on behalf of kids who need help. I probably wouldn't have gone into psychology in the first place if I hadn't stumbled across the path of Dr. Elizabeth Altmaier when I was an undergraduate at the University of Florida (she's now at the University of Iowa).

However, those who were most central to the evolution of many of the ideas in this book, and to whom I owe a tremendous debt of gratitude, were the many kids, parents, educators, and staff with whom I've had the privilege of working over the years. There are truly amazing people in this world who care deeply about improving the lives of kids, have embraced the approach described in this book and, with vision, energy, and relentless determination, have advocated for implementation of the approach in their schools, clinics, inpatient units, and residential and juvenile detention facilities. I have been privileged to cross paths with many of you.

THE
EXPLOSIVE
CHILD

1

THE WAFFLE EPISODE

Saturday morning. Jennifer, age eleven, wakes up, makes her bed, looks around her room to make sure everything is in its place, and heads into the kitchen to make herself breakfast. She peers into the freezer, removes the container of frozen waffles, and counts six waffles. Thinking to herself, "I'll have three waffles this morning and three tomorrow morning," Jennifer toasts her three waffles and sits down to eat.

Moments later, her mother, Debbie, and seven-year-old brother, Riley, enter the kitchen, and Debbie asks Riley what he'd like to eat for breakfast. Riley responds, "Waffles," and Debbie reaches into the freezer for the waffles. Jennifer, who has been listening intently, explodes.

"He can't have the waffles!" Jennifer screams, her face suddenly reddening.

"Why not?" asks Debbie, her voice rising.

"I was going to have those waffles tomorrow morning!" Jennifer screams, jumping out of her chair.

"I'm not telling your brother he can't have waffles!" Debbie yells back.

"He can't have them!" screams Jennifer, now face to face with her mother.

Debbie, wary of the physical and verbal aggression of which her daughter is capable during these moments, desperately asks Riley if there might be something else he would consider eating.

"I want waffles," whimpers Riley, cowering behind his mother.

Jennifer, her frustration and agitation at a peak, pushes Debbie out of the way, seizes the container of frozen waffles, then slams the freezer door shut, grabs her plate of toasted waffles, and stalks to her room. Debbie and Riley begin to cry.

Jennifer's family members have endured literally hundreds of such episodes. In many instances, the episodes are more prolonged and intense and involve more physical or verbal aggression than the one described above (when Jennifer was eight, she kicked out a window of the family car). Doctors have bestowed myriad diagnoses on Jennifer: oppositional-defiant disorder, bipolar disorder, and intermittent explosive disorder. For Jennifer's parents, however, a simple label doesn't begin to capture the upheaval, turmoil, and trauma that her outbursts cause, and doesn't help them understand *why* Jennifer acts the way she does or *when* the outbursts are likely to occur.

Debbie and Riley are scared of her. Jennifer's extreme volatility and inflexibility require constant vigilance and enormous energy from her mother and father, consuming attention the parents wish they could devote to Jennifer's brother. Debbie and her husband, Kevin, frequently argue over the best way to handle her behavior, but agree about the severe strain Jennifer places on their marriage. Jennifer has no close friends; children who initially befriend her eventually find her rigid, bossy personality difficult to tolerate.

Over the years, Debbie and Kevin have sought help from countless mental health professionals, most of whom have urged them to set firmer limits and be more consistent in managing Jennifer's behavior, and have instructed them on how to implement formal reward and punishment strategies, usually in the form of sticker charts and time-outs. When such strategies failed to work, Jennifer was medicated with multiple combinations of drugs, without dramatic effect. After eight years of firmer limits, countless happy faces, and a cornucopia of medicines, Jennifer has changed little since she was a toddler, when Debbie and Kevin first noticed there was something "different" about her. In fact, her outbursts are more intense and more frequent than ever.

"Most people can't imagine how humiliating it is to be scared of your own daughter," says Debbie. "People who don't have a child like Jennifer don't have a clue about what it's like to live like this. Believe me, this is not what I envisioned when I dreamed of having children. This is a nightmare.

"You can't imagine the embarrassment of having Jennifer 'lose it' around people who don't know her. I feel like telling them, 'I have another kid at home who doesn't act like this—I really am a good parent!'

"I know people are thinking, 'What wimpy parents she must have . . . what that kid really needs is a good thrashing.' Believe me, we've tried everything with her. But nobody's been able to tell us how to help her. No one's really been able to tell us what's the matter with her!

"I used to think of myself as a kind, patient, sympathetic person. But Jennifer has caused me to act in ways in which I never thought myself capable. I'm emotionally spent. I can't keep living like this.

"Each time I start to get my hopes up, each time I have a pleasant interaction with Jennifer, I let myself become a little optimistic and start to like her again . . . and then it all comes crashing down with her next outburst. I'm ashamed to say it, but a lot of the time I really don't like her, and I definitely don't like what she's doing to our family. We are in a constant state of crisis.

"I know a lot of other parents whose kids give them a little trouble sometimes . . . you know, like my son. But Jennifer is in a completely different league! It makes me feel very alone."

The truth is, Debbie and Kevin are not alone; there are a lot of Jennifers out there. Their parents often discover that strategies that are usually effective for shaping the behavior of other children—such as explaining, reasoning, redirecting, insisting, reassuring, nurturing, ignoring, rewarding, and punishing—don't achieve the same success with their Jennifers. Even commonly prescribed medications often haven't helped or have made things worse. If you started reading this book because you have a Jennifer of your own, you're probably familiar with how frustrated, confused, angry, bitter, guilty, over-whelmed, spent, scared, and hopeless Jennifer's parents feel.

Clearly, there's something different about the Jennifers of the world. This is a critical realization for parents and others to come to. But there is hope, as long as their parents, teachers, relatives, and therapists are able to come to grips with a second realization: behaviorally challenging kids need us to take a close look at our beliefs about challenging behavior (beliefs most people don't question, unless they're blessed with a behaviorally challenging child) and apply strategies that are often a far cry from ways in which most adults interact with and discipline kids who are not behaviorally challenging.

Dealing more effectively with such kids requires, first and foremost, an *understanding* of why they behave as they do. Once this understanding is achieved, the strategies you'll be reading about in this book will make a lot more sense (and some of the strategies you've been using may start making a lot *less* sense). In some instances, this more accurate understanding can, by itself, lead to improvements in your interactions with your child, even before any formal strategies are tried. Your new understanding of your behaviorally challenging child begins in the next chapter. The new strategies come after that.

Post-meltdown, Debbie sat glumly at the kitchen table, a lukewarm cup of coffee in front of her. Riley was at a friend's house. Jennifer was in her bedroom watching a movie, quiet for now. While Debbie wasn't ecstatic about the amount of time Jennifer spent in front of a screen, she'd come to believe that it was a small price to pay for peace.

Her dilemma: whether to tell Kevin about the waffle incident. Kevin, a high school teacher, had been at the hardware store during the episode. Under normal circumstances, he was a calm, patient man. But he became a completely different person—screaming, threatening—when Jennifer turned family life upside down. He'd never totally lost control, but Debbie was concerned about what he'd do if that ever happened (Kevin had left marks on Jennifer's arms back in the days when they'd tried holding her in time-out. . . she had since convinced him that holding Jennifer in time-out was probably a bad idea).

"I'm not letting that kid rule our lives," Kevin often fumed. Famous last words Debbie thought to herself. If she told Kevin about the waffle episode, she risked having him storm down to Jennifer's room and impose consequences—taking away her DVD player seemed to be his default punishment these days—which would simply ignite another blowout. But if she didn't tell him, Riley probably would, and then Kevin would tell her she was undermining him as an authority figure.

It was during these quiet times that Debbie tended to reflect on Jennifer, who was difficult the instant she came into the world. The nurses at the hospital jokingly forewarned that she and Kevin were in for quite a ride, and Debbie could still picture their smiling faces when they said it. "Freaking hysterical," she now muttered. There were the countless hours spent trying (to no avail) to get Jennifer to stop crying as a baby and toddler. The three preschools that had decided Jennifer was beyond what they could handle. The early calls from the pre-K teachers about other kids not wanting to play with Jennifer because she was bossy and inflexible. There was the suggestion from the kindergarten teachers that Jennifer might benefit from testing or from therapy. There were the play therapists with their toys, the behavioral therapists with their time-outs and sticker charts, the psychiatrists with their medications, the play dates that went badly, the friendship groups Jennifer now refused to attend, the diagnoses, the testing.

But most of all, there were the outbursts.

Their minister urged Debbie to find time for herself. Kevin chuckled when he heard that suggestion: "All you do with your free time is think about Jennifer. You're obsessed." And he was right.

Debbie heard the front door open. "Hello," Kevin called from the front hallway. The hardware store always put him in a good mood.

"In here," called Debbie.

"Got any coffee left?" asked Kevin as he came into the kitchen.

"A little," Debbie said, trying to sound far more chipper than she felt.

Kevin caught the tone in Debbie's voice. "What's the matter?" he said.

"Nothing," said Debbie.

"What'd she do?" asked Kevin.

Here we go, thought Debbie. "Oh, we just had a little incident over waffles while you were gone."

"Waffles?"

"She and Riley had a little dispute over some waffles . . . not a big deal."

"Now she's blowing up over waffles? Geez, what's it gonna be next?"

"Well . . ."

"Where is she?" Kevin's blood was beginning to boil.

"Kevin, I handled it. It's not a big deal. Really. You don't need to do anything."

"Did she hit you?"

"No, she did not hit me. Kevin, it's done."

"You swear she didn't hit you?" Kevin had become aware of his wife's tendency to downplay the severity of the outbursts that occurred when he wasn't around.

"She didn't hit me."

Kevin sighed loudly as he sat down at the kitchen table. Debbie poured him what was left of the coffee.

"Where's Riley?"

"At Stevie's house."

"Did Jennifer hit him?"

"No. Kevin, there was no hitting. Just some screaming. It's really over."

"What's she doing in her room?"

"Watching a video."

"So, as usual, she blows up, and we reward her with a video."

"I've never noticed that depriving her of videos keeps her from blowing up the next time. I just wanted some peace."

"Peace," scoffed Kevin.

Debbie felt tears welling up in her eyes, but blinked them away. "Let's just try to have a nice day."

"In this family, there is no such thing."

2

KIDS DO WELL IF THEY CAN

You know the things that are commonly said about behaviorally challenging kids: they're manipulative, attention-seeking, unmotivated, stubborn, willful, intransigent, bratty, spoiled, controlling, resistant, out of control, and defiant. There's more: they are skilled at testing limits, pushing buttons, coercing adults into giving in, and getting their way. You know (perhaps from personal experience) the things that are said about their parents: they're passive, permissive, inconsistent disciplinarians. They botched the job.

Don't believe any of it. Thanks to the research that's accumulated over the past fifty years or so, we now know better. What we know can be summarized in one sentence:

Behaviorally challenging kids are challenging because they're lacking the skills to not be challenging.

Now, that's a big change in thinking for many people, so let's break it down a little.

Challenging kids are lacking the skills of *flexibility, adaptability, frustration tolerance, and problem solving*, skills most of us take for granted. How can we tell that these kids are lacking those skills? One reason is that the research tells us it's so. But the more important reason is this: *because your child isn't challenging every second of every waking hour.* He's challenging *sometimes*, particularly *in situations where flexibility, adaptability, frustration tolerance, and problem solving are required.* Try to think of the last time your child had an outburst and those skills were *not required.*

Complying with adult directives requires those skills. Interacting adaptively with other people—parents, siblings, teachers, peers, coaches, and teammates—does too. Handling disagreements requires those skills, so does completing a difficult homework assignment or dealing with a change in plan. Most kids are fortunate to have those skills. Your behaviorally challenging child was not so fortunate. Because he's lacking those skills, his life—and yours—is going to be more difficult, at least until you get a handle on things. Understanding *why* your child is challenging is the first step.

Understanding *when* your child is challenging is the second step. Believe it or not, we've already covered that. But let's get more specific:

Challenging behavior occurs when the demands being placed upon a child outstrip the skills he has to respond adaptively to those demands.

When your child has the skills to respond adaptively to demands and expectations, he does. If your child had the skills to handle disagreements and changes in plan and adults setting limits and demands being placed on him without falling apart, he'd be handling these challenges adaptively. Because he doesn't have those skills, he isn't. But let there be no doubt: he'd prefer to be handling those challenges adaptively because *doing well is preferable*. And because—and this is, without question, the most important theme of this entire book—*kids do well if they can*.

So he's not exhibiting challenging behavior because he's enjoying the screaming and shouting and crying and swearing and hitting? No. The kids about whom this book is written do not *choose* to exhibit challenging behavior any more than a child would choose to have a reading disability. They'd prefer to be doing well just like the rest of us. Just like the rest of us, they do poorly when life demands skills they're lacking.

What behaviors does your child exhibit when that happens? Some kids cry, or pout, or sulk, or withdraw. While that's the "easy" end of the spectrum, those kids still need our help. Some hold their breath, scream, swear, kick, hit, have panic attacks, or destroy property. Some run away, bite, cut themselves, vomit, use weapons, or worse. This end of the spectrum is much more concerning and dangerous.

While it's understandable that up until now you've been primarily focused on the *behaviors* your child exhibits when he's upset, those behaviors are actually the least important thing to focus on. As you'll soon see, the most important things to focus on are the *skills* your child is lacking and the specific conditions in which those lagging skills are making life difficult. Believe it or not, those conditions—I call them **unsolved problems**—are actually *highly predictable.* In other words, the belief that challenging episodes occur unpredictably and "out of the blue" (*"We never know what's going to set him off"*) is usually incorrect. Why is this good news? Well, if the unsolved problems setting in motion challenging episodes are predictable, then they can be solved *proactively* rather than in the heat of the moment.

The primary strategy you'll be using to reduce challenging episodes, the strategy this book will be teaching you how to use, is *problem solving.* Not putting stickers on a chart. Not sending your child to time-out (and holding him there when he won't stay). Not screaming. Not berating. Not lecturing. Not sermonizing. Not depriving him of privileges. Not taking away his Xbox for a week. And certainly not spanking. In fact, as you may have noticed, these strategies sometimes *cause* more challenging episodes than they *prevent.*

Let's go back to *kids do well if they can.* This philosophy is important because a different philosophy, the belief that *kids do well if they want to,* has guided adult thinking for a long time. If you believe that your kid isn't doing well because he doesn't want to, then you'll be inclined to use conventional reward and punishment strategies aimed at *making him want to do well.* If you've found that this philosophy and its associated strategies haven't led to a productive outcome with your child, you are not alone. Of course, if that old belief system was serving you well, you wouldn't be reading this book right now.

Throughout this book, I encourage you to put aside the conventional wisdom and strategies and consider the alternate view: that your child is already very motivated to do well and that his challenging episodes reflect a developmental delay in the skills of flexibility, frustration tolerance, and problem solving. *The reason reward and punishment strategies haven't helped is because they won't teach your child the skills he's lacking or solve the problems that are contributing to challenging episodes.* Indeed, you've probably noticed that punishment actually adds fuel to the fire, and that your child only becomes more frustrated when he doesn't receive an anticipated reward. Your energy can be devoted far more productively to collaborating with your child on solutions to the problems that are causing challenging episodes than in sticking with strategies that may actually have made things worse and haven't led to durable improvement.

You also may have noticed that your child's psychiatric diagnosis hasn't provided much information about the skills he's lacking or the specific conditions in which those lagging skills are making life difficult. Diagnoses —such as ADHD, oppositional defiant disorder, bipolar disorder, depression, an autism spectrum disorder, reactive attachment disorder, the newly coined disruptive mood regulation disorder, or any other disorder—can be helpful in some ways. They "validate" that there's something different about your kid, for example. But they can also be counterproductive in that they can cause caregivers to focus more on a child's challenging *behaviors* rather than on the *lagging skills* and *unsolved problems* giving rise to those behaviors. Also, diagnoses suggest that the problem resides within the child and that it's the child who needs to be fixed. The reality is that it takes two to tango. Let there be no doubt, there's something different about your child. But you are part of the mix as well. How you understand and respond to the hand you've been dealt is essential to helping your child.

Now you know that if your kid *could* be more flexible, handle frustration more adaptively, and solve problems more proficiently, he *would*. You also know that lagging skills are getting in his way and contributing to his difficulties. You've read that the unsolved problems that are setting in motion challenging episodes are highly predictable. And now you know that solving those problems together is going be the primary process through which things will get better.

Solving problems together? Yes, indeed. You and your child are going to be allies, not adversaries. Partners, not enemies.

Hard to fathom? It might feel that way right now. But your journey has already begun. We'll get you further down the road in the next chapter.

You have some hard work ahead of you. Though you probably feel like you've been working very hard already, the goal is to make sure you have something to show for that hard work.

Debbie knocked at the door of Jennifer's bedroom, opening it very slightly. "Jennifer, I'm going to take a walk."

Jennifer was still watching a movie, wearing headphones. She didn't acknowledge Debbie's announcement.

Debbie opened the door a bit further—a high-risk move—and raised her voice (another high-risk move). "I'm taking a walk," she yelled.

Jennifer, looking annoyed, paused her movie, and removed one headphone. "Why do you always scream at me?" she groused. But Debbie could sense that, at the moment, Jennifer's level of agitation wasn't extreme.

"I wasn't screaming. I didn't know if you'd heard me."

"I heard you. Can I go to the store later? I need a new pair of rain boots. Mine are too small."

"We can try to do that later, yes," said Debbie.

"Well, can we or can't we?"

"I think we can, but I need to find out what Dad and Riley are doing before I say yes for sure."

"I need rain boots!" Jennifer loudly insisted.

"I know that, Jennifer. I'll do my best."

The ambiguity of Debbie's response had the potential to spark an outburst, but Jennifer was distracted by her movie and the fuse didn't light this time. Debbie was tempted to ask Jennifer what movie she was watching, but decided instead on a quick escape.

Once outside, Debbie called her friend Sandra. They'd met in a support group a few years back and had talked almost daily since. On the surface, they didn't seem to have much in common. Debbie came from a solidly middle-class background, graduated from college, married her high school sweetheart, and had been primed to have the model family (until Jennifer put the kibosh on that game-plan). Sandra came from harder circumstances. She was born to a teenage mom, never knew her biological father, lived with different relatives at various points of childhood and adolescence, was roughed up on several occasions by her mother's boyfriends, ran away and lived on the streets a few times, became pregnant with her son Frankie when she was 16 years old, got her GED at age 20, and was working as an aide at a nursing home and raising Frankie on her own.

Their common bond was behaviorally challenging children. Frankie, whose outbursts were even more severe than Jennifer's, had experienced the "outer edges" of treatment for behaviorally challenging kids. He'd already had multiple placements on inpatient psychiatry units and was in a special education program for kids with emotional and behavioral challenges.

"Hey," Debbie said when Sandra answered. "Do you have a minute?"

"Sure. Just hanging out with Frankie," Sandra said. "He has that flu going around."

"I'm sorry, I'll let you go."

"No, no, he's on the couch watching TV. He's actually rather pleasant when he's sick. Kinda pathetic actually. Makes me wish he was sick all the time."

"You're funny. I feel the same way about Jennifer sometimes. It's the only time she lets me mother her."

"So, what's up?"

Debbie thought it was a little twisted that she looked forward to telling Sandra about Jennifer's latest blow-up—and to hearing Sandra's stories as well—but it made her feel less alone.

"If you can believe it, we had a blowout over waffles this morning."

"Waffles? Why?"

"Well, Jennifer decided that she had a monopoly on the family waffles, and went a little wacko when Riley decided he wanted waffles too."

"Oh my. Was it ugly?"

"A lot less ugly than it could've been. It's actually sort of comical, now that I'm thinking about it . . . watching her stalk off to her room to protect her waffles. Although Riley and I didn't think it was funny at the time. Poor kid."

"That's why I'm glad I don't have any kids besides Frankie. No one has to suffer besides me."

"I feel bad for Riley," said Debbie. "He's gotten a raw deal in the sibling department. But it's nice to have one child who's well-behaved . . . it helps me know I'm actually capable of raising a well-behaved kid."

"Yeah, well, I'm stuck with just my one out-of-control kid. I'm the stereotypical bad single mom. Just ask his teachers."

"I think you deserve a medal for what you've been through with that kid."

"You showing up at my awards ceremony?"

"I think all parents of challenging kids deserve an award," said Debbie. "Not just for what we live with . . . but for tolerating what people say about us."

"Did you tell Kevin?"

"Yeah."

"Did he go nuts?"

"Not this time."

"Are you going to tell Jennifer's therapist about it?"

"I already e-mailed her, but I doubt she'll give me any advice on what to do. She never does. She just meets with Jennifer and they talk about whatever they talk about and I'm left wondering what I'm supposed to do when she goes nuts and my other kid is petrified and my husband loses his mind. Jennifer doesn't even want to go see her anymore. The only reason I make her go is because I need to feel like I'm doing *something*. Otherwise I'd be doing nothing."

"I have a new strategy," said Sandra. "I'm just going to start exposing Frankie to anyone I know who has the flu so he'll stay sick all the time. That's the only thing that seems to work."

3

LAGGING SKILLS AND UNSOLVED PROBLEMS

In the last chapter, you read that lagging skills are *why* kids exhibit challenging behavior more easily, more often, and in more extreme ways than the rest of us do; that challenging behavior occurs when the demands being placed on a child exceed the skills he has to respond adaptively to those demands; that the specific conditions under which that happens are called unsolved problems; and that doing well is always preferable to not doing well. In this chapter, we move beyond the general skills of flexibility, adaptability, frustration tolerance, and problem solving and consider some of the more specific lagging skills that can make it difficult for kids to respond adaptively to life's challenges.

We'll also start being much more specific about unsolved problems.

Because the information presented in this chapter is
really important—especially if you want a deeper understanding of *why* and *when* your child is behaviorally
challenging—I've tried to make the content as engaging as
possible. That said, some folks still find this information
less than electrifying. I encourage you to read it anyway.
When adults understand how lagging skills can set the
stage for challenging episodes, they take the behavior less
personally, respond with greater compassion, and begin
to recognize why what they've been thinking and doing
about their child's challenging episodes may have been
making things worse. When they know what unsolved
problems are setting in motion challenging episodes, they
know exactly what problems need to be solved so those
episodes don't happen anymore.

We'll focus on a sampling of lagging skills first and
then turn our attention to unsolved problems.

LAGGING SKILLS

> *Difficulty handling transitions, shifting from one mind-set or task to an-*
> *other*

Moving from one environment (such as playing outside) to a completely different environment (such as
doing homework inside) requires a shift from one mind-
set (*When I'm playing outside, it's okay to run around and
make noise and socialize*) to another (*When I'm doing*

homework, I need to sit at my desk and concentrate on my schoolwork). If a kid has difficulty with this skill, there's a good chance he'll still be thinking and acting like he's playing outside long after it's time to settle down to do homework. Of course, the situation is likely to become far more precarious when someone is *demanding* that a kid shift mind-sets. This would explain why a kid who lacks this skill could get stuck when, for example, his mother tells him to stop watching television or playing a computer game *immediately* and come to the kitchen for dinner.

That's right, simply *telling a kid what to do* qualifies as a demand for a shift in mind-set. Interestingly, it's when kids are in the midst of having difficulty shifting gears that many adults become even more insistent about instantaneous shifting, which only diminishes the likelihood of efficient shifting and increases the likelihood of a challenging episode.

How do we know a child is having difficulty shifting from one mind-set or task to another? He tells us. Let's listen in:

> **PARENT:** Thomas, it's time to get ready for bed. Turn off the TV please.
>
> **THOMAS:** (No response)
>
> **PARENT:** Thomas, I said it's time to get ready for bed. Turn off the TV.
>
> **THOMAS:** My show's not over yet.
>
> **PARENT:** Yeah, well, your show is never over when it's time for bed. Turn off the TV now please.

THOMAS: My show's over in five minutes! What's the big deal?!

PARENT: The big deal is that I'm tired of you never turning off the TV when I ask you to!

THOMAS: You always tell me to turn off the TV when I'm right in the middle of a show!

PARENT: Thomas, if you don't turn that TV off right now, it will be a very long time before you watch it again!

THOMAS: [*kaboom*]

Does the fact that your kid has difficulty shifting from one mind-set or task to another mean you shouldn't tell him what to do anymore? No, it doesn't mean that. But if telling him what to do heightens the likelihood of a challenging episode, you may want to consider a different approach. Does it mean that you have no expectations? No, you still have expectations. But it would be better for you to rely a lot less on insisting that your child shift gears and imposing consequences when he doesn't, because that doesn't seem to help him shift gears any better. In fact, that sounds like a good recipe for making things worse. If you were hoping that all those outbursts you were enduring when you insisted that your child shift gears would eventually lead to better gear-shifting, now might be a good time to stop hoping. The good news is that your child isn't having difficulty shifting gears all the time; indeed, the shifts he's having trouble making are actually rather predictable (for example, coming in from playing outside, turning the TV off to come in for dinner, ending the video game to get ready for bed, waking up in the morning). So you and your child can come up with a plan for dealing with those situations before they arise again.

> Difficulty considering a range of solutions to a problem
> Difficulty considering past experiences that would guide one's actions in the present
> Difficulty considering the likely outcomes or consequences of one's solutions or potential courses of action (impulsiveness)

What's the main thing your brain must do when you're frustrated? Solve the problem that's causing your frustration. Most of us have never given much thought to the actual thinking processes that are involved in solving a problem because we do it fairly automatically, but if you have a behaviorally challenging kid it's definitely worth thinking about because he's not doing it fairly automatically. First, the process involves identifying the problem you're trying to solve (it's very hard to solve a problem if you don't know what the problem is). Then you need to consider the range of responses or solutions (usually based on past experience) that would help you solve the problem. Next you think about the likely outcomes of each potential solution so as to pick the best one.

Many kids are so disorganized in their thinking that they're unable to figure out what problem they're trying to solve. These kids also have difficulty considering a range of potential solutions and how each solution would pan out. Many are so impulsive that, even if they could think of more than one solution, they've already done the first thing that popped into their heads. The bad news is that, for some kids, the first solution is often the worst one, the one that required the *least* amount of reflection and thought. This probably explains why some behaviorally challenging kids are notorious for putting their worst foot forward. Moreover, there are many behaviorally challenging kids who can't think of any solutions at all. That often causes adults to propose—and often impose—solutions of their own. Unfortunately, the imposition of solutions usually doesn't go over so well, since many behaviorally challenging kids evince a pattern I call *reflexive negativity*: a child's tendency to immediately say "No!" whenever someone proposes a new idea or solution.

Fortunately, behaviorally challenging kids can be helped to approach problems in a more organized, more reflective, less impulsive manner. But imposing solutions isn't the way to do it. That just frustrates your kid further and doesn't help him learn how to think of solutions himself. Sticker charts and time-outs aren't going to get the job done either (nor were they designed to). The good news: because the problems that are giving rise to your child's challenging episodes are actually quite predictable, they can be solved *proactively* and *collaboratively*. And, believe it or not, your kid is probably eager to participate in the process, since he's no more enthusiastic about all the arguing and fighting than you are.

> *Difficulty expressing concerns, needs, or thoughts in words*

Thank goodness humans learned, way back when, how to communicate using words. Language is what separates us from the other species. It's the mechanism by which we exchange information about our thoughts, ideas, concerns, perspectives, and emotions. It's the mechanism by which we *think*. Though most of us haven't thought about it much, language is also the primary way in which we solve problems. Without language, we'd still be solving problems like species whose language processing skills are less developed (barking, biting, running away). Actually, more often than we like to acknowledge, we do stoop to the level of other species, a sign that we humans don't always deploy our skills when we should.

There are many kids whose language processing and communication skills have lagged behind. These kids may not have a basic vocabulary for letting people know they "need a break," that "something's the matter," that they "can't talk about that right now," that they "need a minute" to collect their thoughts or shift gears, or that they "don't like that." Since they lack the wherewithal to adaptively communicate their thoughts, ideas, concerns, perspectives, and emotions, they communicate these things using less optimal words: "screw you," "I hate you," "shut up," and "leave me alone," are some of the milder possibilities. Some kids can't muster any words, so they growl or scream or spit or hit instead. They may have difficulty using internal language (self-talk) to navigate and think their way through potential solutions (*I might not even feel like having waffles tomorrow morning . . . plus, I can ask my mom to buy more today . . . so it's not such a big deal if my brother eats the rest of the waffles right now*). Take Gus, for example:

PARENT: Gus, I understand you got pretty frustrated at school today.

GUS: Yup.

PARENT: What happened?

GUS: Sammy wanted to play with my toys and I didn't want him to.

PARENT: So that made you pretty mad, yes?

GUS: Yes.

PARENT: So what did you do?

GUS: I kicked him.

PARENT: You kicked Sammy?

GUS: Yes.

PARENT: What happened next?

GUS: He told on me.

PARENT: And next?

GUS: I got put in time-out.

PARENT: Did that make you mad?

GUS: Yes.

PARENT: Which part made you mad?

GUS: It made me mad that Sammy took my toys.

PARENT: Were you mad about getting put in time-out?

GUS: Kinda. But I'm in time-out a lot, so I'm kinda used to it.

PARENT: Is it OK for you to kick Sammy?

GUS: No.

PARENT: How come you didn't tell Sammy that you didn't want him to play with the toys you were playing with?

GUS: I didn't know what to say.

PARENT: Is this the first time you and Sammy have had this problem with the toys?

GUS: No, Sammy always wants to play with my toys.

If Gus already knows that he shouldn't kick Sammy, then he doesn't need a time-out to remind him . . . he already knows. Lack of knowledge about behavioral expectations isn't what's getting in his way. If what's really going on is that Gus is having trouble coming up with the words to let Sammy know that he's still playing with certain toys, then we need to help him solve that problem (something no amount of time-outs will accomplish). As long as Gus doesn't have the words, he's going to keep kicking Sammy. If, as Gus suggests, this isn't the first time that Gus and Sammy have had a conflict over sharing toys, then this is a highly predictable unsolved problem and it can be solved proactively. Even if the problem has never arisen before, it's predictable now (because it's happened once). Doesn't Sammy need to know that adults are taking the problem seriously? Yes, and giving Gus time-outs wouldn't be the best way to do it. Solving the problem in a way that helps Gus and Sammy figure out how to share toys and that helps Gus learn some new words that he can use in such situations (so he's not kicking instead) would be a very good way to do it.

> *Difficulty managing emotional response to frustration in order to think rationally*

As you know, solving problems is much easier if a person has the ability to think through solutions. The emotions people feel in the midst of frustration can make rational thinking more difficult. It's not that the emotions are all bad: they can be useful for mobilizing or energizing people to solve a problem. But the skill of putting one's emotions aside so as to think through solutions to problems more objectively, rationally, and logically is really important. Kids who are pretty good at this skill tend to respond to problems or frustrations with more thought than emotion, and that's good. Children whose skills in this domain are lacking tend to respond to problems or frustrations with less thought and more emotion, and that's not good at all. They may actually feel themselves "heating up" but often are unable to stem the emotional tide until later, when the emotions have subsided and rational thought has kicked back in. Then they're often remorseful for what happened when they were upset. They may even have the knowledge to deal successfully with problems and can actually demonstrate such knowledge under calmer circumstances. But when they're frustrated, their powerful emotions prevent them from accessing and using what they know. Of course, some kids aren't very good at the thinking part either, so they deal with problems only at an emotional level. And emotions don't solve problems. You know what this looks like:

PARENT: Philip, come eat the scrambled eggs I made for breakfast.
PHILIP (RESPONDING WITH MORE EMOTION THAN THOUGHT, BUT ALSO TELLING THE TRUTH): I hate scrambled eggs! You always make things I don't like!
PARENT: Well, that's what I made your sister! I made enough for both of you!
PHILIP: Well that's not what I want!

PARENT: I'm not running a restaurant! And I'm not sending you to school with an empty stomach! Eat the eggs!
PHILIP (DUMPING THE EGGS IN THE SINK): No, I hate eggs!
PARENT (NOW PERHAPS ALSO RESPONDING WITH MORE EMOTION THAN THOUGHT): Your Xbox is history, pal!
PHILIP: [*kaboom*]

If you respond to a child who's having difficulty putting his emotions aside so as to think through solutions by imposing your will more intensively and "teaching him who's the boss," you probably won't help him manage his emotions. Quite the opposite, in fact. But the dialogue above suggests that Philip's unsolved problem—difficulty eating what mom has made for breakfast—is actually highly predictable, which means it can be solved ahead of time.

So far we've only talked about in-the-moment emotion regulation. But there are some kids whose difficulties in managing emotions are more chronic. These kids are irritable, agitated, cranky, and fatigued much more often and much more intensely than others. Most of us have more trouble handling frustration and solving problems when we're in a bad mood. But these kids are in a bad mood a lot, so they have trouble handling frustration and solving problems a lot, too:

MOTHER: Mickey, why so grumpy? It's a beautiful day outside. Why are you indoors?
MICKEY (SLUMPED IN A CHAIR, AGITATED): It's windy.
MOTHER: It's windy?
MICKEY (MORE AGITATED): I said it's windy! I hate wind!

MOTHER: Mickey, you could be out playing basketball, swimming . . . you're this upset over a little wind?

MICKEY (VERY AGITATED): It's too windy, damn it! Leave me alone!

MOTHER: Should we try to think of something you could do instead?

MICKEY: There's nothing else to do instead.

Because it can get in the way of rational thought, anxiety can have the same effect as irritability. When a kid is anxious about something—a monster under the bed, an upcoming test, a new or unpredictable situation—clear thinking is essential. A little anxiety can actually be helpful, because it can spur a person to take action. But too much anxiety can make rational thinking much harder, which only makes the person more anxious.

If a kid is already upset, threats and imposed solutions and time-outs often simply fuel the fire. And if he's not already upset, threats and imposed solutions and time-outs are a good way to get him there. So we're going to need a different approach to solving those problems. Different timing, too.

> *Difficulty seeing the "grays;" concrete, literal, black-and-white thinking*
> *Difficulty deviating from rules or routine*
> *Difficulty handling unpredictability, ambiguity, uncertainty, or novelty*
> *Difficulty shifting from an original idea or solution*
> *Difficulty adapting to changes in plan or new rules*
> *Difficulty taking into account situational factors that would suggest the need to change a plan*

Very young children tend to be fairly rigid, black-and-white, literal thinkers. That's because they're still making sense of the world, and it's easier to put two and two together if you don't have to worry about exceptions to the rule or alternative ways of looking at things. As children develop, they learn that, in fact, most things in life are "gray"; there *are* exceptions to the rules and alternative ways of interpreting things. We don't go home from Grandma's house the same way every time; we don't eat dinner at the exact same time every day; and the weather doesn't always cooperate with our plans.

Unfortunately, for some children, "gray" thinking doesn't develop readily. These kids sometimes end up with diagnoses on the autism spectrum. But regardless of diagnosis they're best thought of as *black-and-white thinkers living in a gray world*. They have significant difficulty approaching the world in a flexible, adaptable way and become extremely frustrated when events don't proceed in the manner they had anticipated. More specifically, these children have a strong preference for predictability and routines, and struggle when events are unpredictable, uncertain, or ambiguous.

Such kids run into trouble when they need to adjust or reconfigure their expectations. They tend to over-focus on facts and details and often have trouble recognizing the obvious or seeing the big picture. For example, a child may insist on going out for recess at a certain time on a given day because that's the time the class always goes out for recess, failing to take into account both the likely consequences (e.g., being at recess alone) of insisting on the original plan of action and important situational factors (an unexpected assembly, perhaps) that would suggest the need for a change in plan. These children may experience enormous frustration as they struggle to apply concrete rules to a world where few such rules apply:

> **PARENT:** Courtney, we can't go to the park today . . . it's raining.
>
> **COURTNEY:** But we were supposed to go to the park!
>
> **PARENT:** I know. I wish it wasn't raining, but I don't really see how we can still go. We'd get all wet.
>
> **COURTNEY:** No, we still have to go to the park! That's the plan!
>
> **PARENT:** We can always go tomorrow, if the weather's nicer.
>
> **COURTNEY:** We're supposed to go today!
>
> **PARENT:** How 'bout we go to a movie instead?
>
> **COURTNEY:** No! We're supposed to go to the park!
>
> **PARENT:** Look, Courtney, it's raining. We'll get all wet. I'm not going to the park in the pouring rain!
>
> **COURTNEY (RUNNING TOWARD THE DOOR):** I'm going to the park!
>
> **PARENT (BLOCKING COURTNEY FROM THE DOOR):** You can't go to the park!
>
> **COURTNEY:** [*kaboom*]

After the storm passes, the parent might try asking the usual:

PARENT: Courtney, how come you got so upset when we couldn't go to the park because of the rain?
COURTNEY: I don't know.

That's actually a pretty informative response, though it may not seem like it. In a perfect world, the child would respond by saying something like this:

"See, guys, I have a little problem. Actually, it's turning into a big problem. I'm not very good at being flexible, handling frustration, and solving problems. And you—and lots of other people—expect me to handle changes in plans, being told what to do, and things not going the way I thought they would as well as other kids. When you expect these things, I start to get frustrated, and then I have trouble thinking clearly, and then I get even more frustrated. Then you guys get frustrated, and that just makes things worse. Then I start doing things I wish I didn't do and saying things I wish I didn't say. Then you sometimes do things you wish you didn't do and say things you wish you didn't say. Then you punish me, and it gets really messy. After the dust settles— you know, when I start thinking clearly again—I end up being really sorry for the things I did and said. I know this isn't fun for you, but rest assured, I'm not having any fun either."

Behaviorally challenging kids are rarely able to describe their difficulties with this kind of clarity. But here's a simple math equation that might suffice.

Inflexibility + Inflexibility = Meltdown

Fortunately, children like Courtney can be helped to approach the world in a grayer, more flexible manner and can participate in solving the problems that are causing their challenging episodes. Of course, we wouldn't want to wait until we're in the midst of yet another challenging episode to try solving those problems. We want to solve them proactively.

Our overview of lagging skills is now complete. Of course, that was just a sampling. Here's a more complete, though hardly exhaustive, list, including those we just reviewed:

> *Difficulty handling transitions, shifting from one mind-set or task to another*
> *Difficulty doing things in a logical sequence or prescribed order*
> *Difficulty persisting on challenging or tedious tasks*
> *Poor sense of time*
> *Difficulty maintaining focus*
> *Difficulty considering the likely outcomes or consequences of actions (impulsive)*
> *Difficulty considering a range of solutions to a problem*
> *Difficulty expressing concerns, needs, or thoughts in words*
> *Difficulty understanding what is being said*
> *Difficulty managing emotional response to frustration so as to think rationally*
> *Chronic irritability and/or anxiety significantly impede capacity for problem-solving or heighten frustration*
> *Difficulty seeing the "grays"/concrete, literal, black-and-white thinking*
> *Difficulty deviating from rules, routine*
> *Difficulty handling unpredictability, ambiguity, uncertainty, novelty*

> *Difficulty shifting from original idea, plan, or solution*
> *Difficulty taking into account situational factors that would suggest the need to adjust a plan of action*
> *Inflexible, inaccurate interpretations/cognitive distortions or biases (e.g., "Everyone's out to get me," "Nobody likes me," "You always blame me," "It's not fair," "I'm stupid")*
> *Difficulty attending to or accurately interpreting social cues/poor perception of social nuances*
> *Difficulty starting conversations, entering groups, connecting with people/lacking basic social skills*
> *Difficulty seeking attention in appropriate ways*
> *Difficulty appreciating how his/her behavior is affecting other people*
> *Difficulty empathizing with others, appreciating another person's perspective or point of view*
> *Difficulty appreciating how s/he is coming across or being perceived by others*
> *Sensory/motor difficulties*

If after reading this list you've concluded that your behaviorally challenging child has quite a few lagging skills, that's good. Not good that your child is lacking all those skills, but definitely a good thing that you now know about them. Those lagging skills can take the place of all the other things that have been said about your child, including the following:

He just wants attention.
This common cliché is often invoked to explain why kids are behaviorally challenging . . . but, since we *all* want attention, it doesn't help us understand what's really get-

ting in a child's way, and it doesn't answer the more critical questions: If the kid has the skills to seek attention adaptively, then why is he seeking attention in such a maladaptive fashion? Doesn't the fact that he's seeking attention maladaptively tell us he doesn't have the skills to seek attention adaptively?

He just wants his own way.

We all want our own way; some of us have the skills to get our own way adaptively, and some of us don't. This cliché doesn't help us understand why a kid is going about getting his own way in such a maladaptive fashion. Adaptively getting one's own way requires a lot of skills often found lacking in behaviorally challenging kids.

He's manipulating us.

This is another popular but misguided way of portraying behaviorally challenging kids. Competent manipulation requires various skills—forethought, planning, impulse control, organization—that behaviorally challenging kids often lack.

He's not motivated.

If it's true that *kids do well if they can*, then the kid is already motivated and needs something else from us besides rewards and punishments. Remember, if the kid *could* do well he *would* do well, so poor motivation is unlikely to be what is truly keeping him from doing well.

Rewards and punishments don't teach lagging thinking skills and don't solve the problems that precipitate challenging episodes.

He's making bad choices.

This suggests that the kid already has the skills to be making good choices. Of course, if he had those skills, we wouldn't be wondering why he's making so many bad choices!

He has a bad attitude.

He probably didn't start out with one. "Bad attitudes" tend to be the by-product of countless years of being misunderstood, over-corrected, over-directed, and over-punished by adults who didn't recognize that a kid lacked crucial thinking skills. But kids are resilient; they come around if we start doing the right thing.

He knows just what buttons to push.

We should reword this one so it's more accurate: *when he's having difficulty being flexible, dealing adaptively with frustration, and solving problems, he does things that are very maladaptive and that adults experience as being extremely unpleasant.*

He has a mental illness.

I'm not sure what this means anymore. If it simply means that a kid qualifies for a psychiatric diagnosis, then, at the risk of redundancy, I should point out that with a diagnosis we still don't know what skills he's lacking or what unsolved problems are reliably and predictably precipitating his challenging episodes. I prefer *problems in living* (a term originated a long time ago by Dr. Thomas Szasz) to the term *mental illness*, because it highlights what behaviorally challenging kids actually need: help solving the problems that are setting the stage for their challenging episodes.

Many adults new to these ideas tell me they feel guilty, both about what they didn't know and about how they've been treating their child. Why didn't any of the mental health professionals you've seen previously tell you about lagging skills? Probably because their training took them in a different direction. As for the guilt, I understand. But you didn't know what you didn't know. Now you know: these skills don't come naturally to all children. We tend to think that all children are created equal in these capacities, and this assumption causes many adults to believe that behaviorally challenging children must not *want* to do well. Now you know better.

By the way, there's a big difference between interpreting lagging skills as *excuses* rather than as *explanations*. When lagging skills are invoked as excuses, the door slams shut on the process of thinking about how to help a child. But when lagging skills are used as explanations for a child's behavior, the door swings wide open to a vast array of alternative options for helping him.

UNSOLVED PROBLEMS

Let's now turn our attention to *when* your child is exhibiting challenging behavior. Don't let the fact that this section is at the end of the chapter fool you; when it comes to reducing challenging episodes, unsolved problems are even more important than lagging skills.

As you've read, behaviorally challenging kids aren't challenging every second of every waking hour. They're challenging *sometimes*: in the specific situations in which the demands or expectations being placed upon them outstrip the skills they have to respond adaptively. As you know, those situations are called unsolved problems. How are we going to dramatically reduce challenging episodes? We're going to solve those problems.

But first we have to figure out what those problems are. Here are some examples (there are a lot more in the next chapter). Let's start with homework, which seems to be the number one problem triggering challenging episodes in North American households. If homework completion reliably and predictably precipitates challenging episodes, then homework completion is a problem you and your child will need to solve (as you'll read in the next chapter, you'll want to be precise about the specific homework assignments that your child is having difficulty completing). If there are chores your child is having difficulty completing and that sets off challenging episodes, then completing chores is a problem you and your child will need to solve (again, you'll want to be very specific about the chores your child is having difficulty completing). If you feel that your child is spending an inordinate amount of time in front of a screen (TV, computer, video game) and that the screen time is interfering with other aspects of living, and your efforts to separate your child from the screen is setting in motion challenging episodes, then that's a problem you and your child will need to solve. (Once again, you'll want to be specific about the aspects of living, such as eating dinner with the family or getting ready for bed, that the screen time is making difficult.)

At the end of the next chapter, you're going to be assigned the task of identifying your kid's lagging skills and unsolved problems. Once you identify the unsolved problems that are precipitating challenging episodes, those episodes become *highly predictable*. Most behaviorally challenging kids are reliably set off by the same five or six (or ten or twelve) problems every day or every week. As you know, many people believe that a kid's challenging episodes are unpredictable and occur "out of the blue." That may explain why they wait until a problem "pops up" (yet again) before they try to deal with it (yet again). That's seldom an effective or reliable strategy. Luckily, because those episodes and the problems setting them in motion are predictable (in other words, they don't really "pop up"), they can be solved proactively. A very important goal of this book is to help you begin solving problems proactively rather than emergently and reactively. You want to be in crisis *prevention* mode, not crisis *management* mode.

One more very important thing before this chapter ends: it's not just the *timing* on your problem solving that will probably need some adjustment, it's how you're going about trying to solve the problems in the first place.

Most adults favor *unilateral* problem solving: they decide on the solution, impose that solution on the child, and provide the child with an incentive—rewards and punishments—for adhering to that solution. That's how most of us were raised. But that approach to solving problems is also what often *causes* your child's challenging episodes (you'll read more about this in Chapter 5). That's going to need some adjusting as well. Beginning with Chapter 6, this book is devoted to helping you learn how to solve problems *collaboratively*, rather than unilaterally.

If you start solving problems *collaboratively* and *proactively*, you're going start seeing a dramatic reduction in challenging episodes. Yes, there will be many bumps in the road (learning to solve problems collaboratively and proactively is hard). But, just by reading these first three chapters, you've made a lot of progress already.

YOU

Why would a chapter describing lagging skills and unsolved problems end with *you*? Well, your kid is only half of the equation. We also need to consider the other half: what expectations you're placing on your kid, whether those expectations are truly realistic, how you're going about trying to get your expectations met, and what problem-solving approach you're using when your expectations are not met. As you well know, the frustration you experience in trying to understand and interact with your behaviorally challenging child can, at times, have *you* behaving at your worst. For things to improve, your kid is going to need you to be at your best. It's the goal of this book to get you there. Thus, as you continue reading, *you* are going to be an increasingly important part of the picture. Challenging episodes don't occur in a vacuum. It takes two to tango.

Here are this chapter's key points:

- There are various lagging skills that can make it difficult for a kid to respond to life's challenges in an adaptive, rational manner.

- One of the biggest favors you can do for a behaviorally challenging kid is to identify the lagging skills that are contributing to challenging behavior so that you and others understand what's getting in his way.

- The other big favor you can do for a challenging kid is to identify the specific unsolved problems that are reliably and predictably precipitating his challenging episodes. Once those unsolved problems are identified, challenging episodes become highly predictable.

- We're going to significantly reduce challenging episodes in your household by changing the way you go about trying to solve those problems with your kid.

You can find a form on which you can identify all of your child's lagging skills and unsolved problems—called the *Assessment of Lagging Skills and Unsolved Problems*—at www. livesinthebalance.org (it's in a section called The Paperwork).

4

GETTING
STARTED

As you now know, your journey begins by identifying your child's lagging skills and unsolved problems. In this chapter, we're going to discuss how to do that, using the *Assessment of Lagging Skills and Unsolved Problems*. If you haven't downloaded it yet, please see page 42.

Here's how you use it. You start with the first lagging skill and consider whether it applies to your child. If it doesn't apply, move on to the next lagging skill. If it does apply to your child, check it off and, instead of moving on to the next lagging skill, move over and begin identifying the unsolved problems that seem to be associated with that lagging skill. (I don't recommend going through the entire list of lagging skills first and then going back to identify unsolved problems.) After you've identified one unsolved problem that seems to be associated with a particular lagging skill, identify some more. In other words, identify all the unsolved problems connected to that lagging skill before moving on to the next lagging skill.

Identifying (and checking off) your child's lagging skills isn't the hard part. Identifying the unsolved problems associated with each lagging skill tends to be a little harder, and there are a few guidelines you'll want to keep in mind. The guidelines aren't there to make things harder (though it may seem that way at first); they're there to make it more likely that your child will participate in the problem-solving process.

GUIDELINE #1: *Whenever possible, start each unsolved problem with the word "difficulty."*

For example, "Difficulty taking the trash out in the morning," or "Difficulty completing the paragraphs on the Language Arts homework," or "Difficulty getting dressed before school in the morning." The unsolved problem shouldn't include the behaviors your child is exhibiting in response to the unsolved problem. The word *difficulty* covers those behaviors. In other words, you wouldn't write "Screaming and crying when having difficulty taking the trash out in the morning."

Why is it important to leave the behavior out of the unsolved problem? Because your wording of the unsolved problem on the ALSUP is going to translate directly into how you introduce the unsolved problem to your child when you're beginning the problem-solving process. Many kids become defensive and won't participate in the process if you highlight their challenging behavior as you're trying to get the conversation rolling. So there's no need to include the challenging behavior in the unsolved problem; the word "Difficulty" is more neutral.

GUIDELINE #2: *Make sure the unsolved problems are sufficiently specific and "split" rather than "clumped."*

Here's an example of a *clumped* unsolved problem: *Difficulty completing homework.* What's wrong with that wording? If your child is having difficulty completing *many* different homework assignments, clumping them together will make it more difficult for him to provide information about what's hard about each specific assignment. If your child is having difficulty writing the paragraphs on the Language Arts homework as well as with completing the homework involving double-digit division problems, those should be treated as two separate unsolved problems, even though they're both related to homework. And if he's also having difficulty memorizing his multiplication tables, then that's a separate unsolved problem too, even though it's also related to math. We don't want to assume that your child is having difficulty on those unsolved problems for the same reasons.

Are "splitting" and being specific going to make your list of unsolved problems longer? Absolutely. But let's not kid ourselves: your list of unsolved problems *is* very long. Thank goodness you're now identifying those unsolved problems and being very specific about them, so you can start solving them.

GUIDELINE #3: *Keep your theories about the cause of the unsolved problem out of the unsolved problem.*

You wouldn't write "Difficulty completing the paragraphs on the Language Arts homework because she just doesn't feel like doing them" because "she just doesn't feel like doing them" is your theory. Follow this rule of thumb: the minute you're inclined to write the word *because* in the unsolved problem, stop writing. Everything that comes after *because* is a theory. See, there are a few pitfalls with adult theories. First of all, adult assumptions and theories about the cause of an unsolved problem are often incorrect. Second, including a theory in an unsolved problem can make it harder for your kid to tell you his actual concern or perspective, and could unnecessarily inflame him as well. Don't worry. You'll find out what's really going on when you start trying to solve the problem with your child and, under most circumstances, it's your child who's going to provide that information. The hardest thing about this guideline is that adults love their theories. You'll need to fall out of love with your theories if you want to find out what's really getting in your child's way on the unsolved problems that have been causing challenging episodes.

GUIDELINE #4: *Keep solutions out of the unsolved problems.*

Here's an example of an unsolved problem that includes a solution: *Difficulty laying out the clothes you're going to wear to school the night before so you don't have difficulty getting dressed for school in the morning.* The unsolved problem here is that the child is having difficulty getting dressed for school in the morning. Apparently, a solution to this problem is already being applied (laying out clothes the night before). If that solution were working, this wouldn't still be an unsolved problem. We can forego the solution in the wording of the unsolved problem and simply write "Difficulty getting dressed for school in the morning."

A few nights later, after Jennifer and Riley were in bed, Debbie and Kevin sat down together at the kitchen table. Debbie had printed out two copies of the *Assessment of Lagging Skills and Unsolved Problems* (ALSUP). Their goal: to identify Jennifer's lagging skills and unsolved problems.

"Now, what we're supposed to do is start at the top of the list of lagging skills and decide if the first lagging skill applies to Jennifer," explained Debbie. "If a lagging skill applies, we check it off and then move over and identify the unsolved problems that are associated with that lagging skill."

Kevin began scanning his copy of the ALSUP. "Why are we doing this?"

"Because after all these years, we still don't know why Jennifer is the way she is," said Debbie.

Kevin sighed. "And we can figure this out on our own?"

"It's not like any of the doctors we've seen have nailed it," said Debbie.

"And this sheet of paper is going to tell us?"

"Yes," said Debbie, "and by the way, I already looked at the list of lagging skills, and she basically lights up the board."

"You started without me?" said Kevin, feigning insult.

"I start everything without you," Debbie smiled.

"OK, let's go. But I don't know what an unsolved problem is."

"Unsolved problems are the situations in which Jennifer's getting upset."

Kevin was a little confused again. "Where do we put stuff like hitting, and screaming, and swearing?"

"We don't," said Debbie. "Those are the things she's doing *because* of the unsolved problems, but they aren't the unsolved problems."

"Hitting seems like a pretty big problem to me," said Kevin.

"Yeah, but that's not what we're going to be working on with Jennifer," said Debbie. "That's the whole point. All these years we've been focused on her *behavior*, when we should have been focused on solving the problems that *cause* her behavior."

"I'm not going to let her hit people," said Kevin.

"Yes, I know we're not going to let her hit people," said Debbie, trying to stay patient. "But we're going to get rid of the hitting by solving the problems that are causing her to get upset."

"Jennifer's always getting upset," said Kevin.

"Yeah, that's what I've always thought," said Debbie. "But she's not always getting upset. We have to be more specific about when she's getting upset. Otherwise, we won't know what problems we're trying to solve with her. How about we just try one? What about this first lagging skill? 'Difficulty making transitions.' What do you think?"

"I think that one's true," said Kevin.

"I agree," said Debbie. "So I'm going to check it off. Now we have to figure out *when* Jennifer's having trouble making transitions. Those are the unsolved problems."

"I get it," said Kevin. "Like when we're trying to get her to turn off her video and go to bed?"

"Yes, that's a good example. So what we do now is we write that down in the unsolved problems section. Unsolved problems should begin with the word 'difficulty.'" She wrote *"Difficulty turning off the video to get ready for bed at night"* in the unsolved problems section. "Let's think of some more examples."

"Well, any time we want her to turn off her video to do anything."

"Right," said Debbie. "But we're supposed to 'split' the problems instead of 'clumping' them together. So we need to name all the specific times she has trouble making transitions."

"OK . . . like when we want her to turn off the video and come with us to church."

"That's a good one too," said Debbie, writing on the ALSUP.

"This isn't so hard," said Kevin.

"Nope, not hard at all. We should have done this ten years ago."

"She's going to have a lot of unsolved problems," Kevin observed.

"Of course she has a lot of unsolved problems!" said Debbie. "And all this time, we could have been busy solving them! But that's not what we've been doing. We've been getting her diagnosed, and giving her stickers, and yelling at her, and getting hit. We've been spinning our wheels!"

"OK, OK. But how are we going to solve all the problems once we know what they are?"

"Let's stay focused here," said Debbie. "That comes next."

"Are you sure we don't need a doctor to help us with this?" Kevin wondered.

"I don't know if we need a doctor to help us with this," said Debbie. "We seem to be doing fine right now. Can we keep going please?"

To aid in the digestion of the guidelines—and to provide many specific examples of unsolved problems— let's see what else Jennifer's parents, and then Frankie's mom, came up with. Then we'll turn our attention to some other behaviorally challenging kids. Then it's your turn.

JENNIFER

Here's the complete list of lagging skills Jennifer's parents felt applied to their daughter:

> Difficulty handling transitions, shifting from one mind-set or task to another
> Difficulty considering the likely outcomes or consequences of actions (impulsive)
> Difficulty considering a range of solutions to a problem
> Difficulty managing emotional response to frustration so as to think rationally
> Chronic irritability and/or anxiety significantly impede capacity for

problem-solving or heighten frustration
> *Difficulty seeing the "grays"/concrete, literal, black-and-white thinking*
> *Difficulty deviating from rules, routine*
> *Difficulty handling unpredictability, ambiguity, uncertainty, novelty*
> *Difficulty shifting from original idea, plan, or solution*
> *Difficulty taking into account situational factors that would suggest the need to adjust a plan of action*
> *Difficulty attending to or accurately interpreting social cues/poor perception of social nuances*
> *Difficulty starting conversations, entering groups, connecting with people/lacking basic social skills*
> *Difficulty appreciating how his/her behavior is affecting other people*
> *Difficulty empathizing with others, appreciating another person's perspective or point of view*
> *Difficulty appreciating how s/he is coming across or being perceived by others*

Here's a *partial* list of the unsolved problems:

> *Difficulty when the food she wants for breakfast (usually waffles) isn't available.*
> *Difficulty when something she wants to wear is still in the laundry.*
> *Difficulty eating at the dinner table with Mom, Dad, and Riley.*
> *Difficulty when Mom has packed something for her lunch that she doesn't like.*
> *Difficulty turning off a video to get ready for bed at night.*
> *Difficulty turning off a video when it's time to go to church.*
> *Difficulty when she wants to buy something (for example, rain boots), but Mom or Dad can't do it right away.*

> *Difficulty agreeing with Riley on what to watch on TV.*
> *Difficulty when Riley doesn't want to play with her.*
> *Difficulty agreeing on what restaurant to eat at when the family goes out to to dinner.*
> *Difficulty being in bed with the lights out by 10 pm.*
> *Difficulty going to the park with Mom, Dad, and Riley.*
> *Difficulty going to Grandma's house with Mom, Dad, and Riley.*
> *Difficulty finding a friend to hang out with on weekends.*
> *Difficulty if something has been moved in her room.*

Remember, even though lagging skills and unsolved problems are being listed separately here, you're actually discussing and identifying them simultaneously. If a lagging skill applies to your child, you're checking it off and then identifying the unsolved problems that are associated with that lagging skill before moving on to the next lagging skill. Also—I know I'm being a little redundant here, but this is important—notice that "hitting" and "screaming" and "swearing" are not in the list of unsolved problems. That's because those are *behaviors* and the word *difficulty* has taken their place. It's far more important to be specific about the unsolved problems causing the challenging behaviors than it is to be specific about the behaviors that are associated with those unsolved problems.

Kevin's mood became increasingly somber as he and Debbie completed the ALSUP. "This is kind of sad that we didn't know this stuff about our own daughter."

"Sad for us and sad for her," said Debbie.

"And don't forget about Riley," said Kevin. "How come we didn't figure this stuff out before?"

"Because we didn't know what we didn't know," said Debbie.

"So all those diagnoses she has . . . they don't mean anything?"

"They certainly didn't help me understand Jennifer as well as the lagging skills and unsolved problems do," said Debbie.

"I'm a teacher," said Kevin. "I should know this stuff."

"I don't think most teachers know about this stuff," said Debbie.

"So we've been doing it wrong all this time?"

"Well, we've been doing what we were told to do," said Debbie. "Maybe what they've been telling us to do works for some kids. But it didn't work for ours."

FRANKIE

Frankie was playing a video game in his bedroom. Sandra paced in the living room, smoldering. After missing several days of school because of the flu, Frankie had been suspended that day—his first day back—for swearing at a teacher.

They told me they knew how to handle kids like Frankie, Sandra fumed. *He promised to try as hard as he could to stay out of trouble. Now he's blowing it, again. And after only a month!*

Anger had been a familiar companion since Sandra was a kid. Back then, it wasn't just her circumstances that fueled her anger, it was also the fact that she could do very little to change those circumstances. The anger had always energized her to fight harder. But her anger and determination always seemed to backfire in her interactions with Frankie; they just caused him to fight back.

His new school program had given her hope that maybe he would turn a corner. Now this. As she was in the midst of getting her bearings on whether to cry or scream, her phone rang.

It was Debbie, who immediately sensed Sandra's agitation.

"Are you OK?" asked Debbie.

"I'm trying to decide who to scream at," Sandra replied.

"What happened?"

"Frankie got suspended from school today."

"I'm sorry. I guess he's over the flu."

"Oh, he seems to be back in full force."

"Did you have to miss work?"

"I had to leave work to pick him up. My boss said she can't keep letting me do that." Sandra tried to keep her voice from trembling. "How the heck am I supposed to make this work?"

"I'm sorry," said Debbie.

"It's a freaking special education program! They're supposed to be able to handle him! What are they sending him home for?"

"It makes no sense," said Debbie, trying to empathize.

"And I don't even know what happened! All I know is he swore at someone. What am I supposed to do about it?! I wasn't there! This always happens. He does well for a while, then he screws it up."

"What are you going to do?"

"They want me to come to a meeting tomorrow. I have to miss work for that too. I swear he's going to get me fired. Then we won't have a place to live either."

"I'm sorry," Debbie said again.

Sandra took a deep breath, exhaling slowly. "So help me decide who to scream at," she said, only half joking.

"Who are the candidates?" asked Debbie.

"My son, for starters. But if I scream at him he'll scream back and it'll get ugly, and that never accomplishes anything."

"OK, so cross him off your list. Who else?"

"Do we have to cross him off the list so fast?" asked Sandra, still only half joking. "I cannot tell you how tired I am of dealing with this crap."

"I know," said Debbie. "I guess you could scream at him if you really want to. But I don't think the ugly part is going to thrill you. Who else could you scream at?"

"The director of his program at school. But that's pointless too. He'll just think I'm a crazy mom who's over-protective of her kid. Been there, done that."

"Probably worth crossing him off your list too. Any others?"

"That's pretty much it. Guess I'm not screaming at anybody."

"You can scream at me a little if it would help."

Sandra laughed. "I think I did that already. Sorry."

"You've been dealing with things going badly at school for a long time." Debbie found herself thinking about what she'd learned from the website on which she'd found the ALSUP. "Too bad they're not focused on his lagging skills and unsolved problems."

"His what?"

"I found this website . . ." Debbie paused. "You know what, this might not be the best time."

"Lay it on me, honey. I need a diversion."

"Are you sure you want to hear about it right now?"

"No time like the present. Is it going to keep my kid in school?"

"Um, I don't know. But I thought it was pretty informative. It helped Kevin and me learn things about Jennifer that we didn't know."

"Oh, I think I know Frankie pretty well. He hits, he screams, he swears, he gets thrown out of school . . ."

"Well, that's just it. According to this website, those behaviors aren't the most important thing about Frankie," said Debbie. "Those are his behaviors."

"They seem pretty important to the people who are throwing him out of school," said Sandra, unconvinced.

"I know. But the important part is *why* he's doing those things."

"He's doing those things because he's bipolar. You know that," said Sandra.

"Bipolar disorder is just his diagnosis," said Debbie, "but it's not why he does that stuff."

"And this website is going to tell me why?"

"Yeah, and what to do about it," said Debbie.

"Oh, I don't think there's anything anybody can do about it," said Sandra, even more dubious. "He's been on every medicine known to mankind. He's pretty severe."

"Yeah, but part of the reason he's pretty severe is because no one's ever figured out what's really going on with him," said Debbie. "There's a form on the website that's really eye-opening. It helped us understand Jennifer better, and it helped us nail down the problems that get her upset."

"Geez, you're really pumped up."

"All I know is, it's the first time Kevin and I have been on the same page about anything related to Jennifer. It makes it obvious why all the stuff we've been doing hasn't worked. We've been focused on the wrong things! You should really check it out!"

"OK, I'll check it out. But I'm not getting my hopes up," said Sandra.

Debbie paused. "You've been through a lot. OK, so don't get your hopes up. But check out the website."

Sandra was too tired and angry to do anything that night. But a few nights later, she looked at the ALSUP on the website. She became a little confused when she tried to identify Frankie's unsolved problems and called Debbie.

"I tried filling out the ALSUP," said Sandra.

"Good for you!" said Debbie. "Well?"

"I didn't get very far."

"How come?"

"Well, the lagging skills were easy enough to check off. But I couldn't quite figure out the unsolved problems."

"We had trouble with that, too. It's easier to think of the behaviors than the problems that are causing them."

"Exactly! How'd you do it?"

"Well, every time I thought of a behavior, I thought of the situations in which the behavior occurs. The situation is the unsolved problem."

"So, like screaming. Frankie screams all the time."

"What's he screaming about?" asked Debbie.

"Everything."

"Yeah, but *what* exactly? What's an example of something he screams about?"

"He's screaming because I want him to turn down the volume on his music."

"He doesn't have earbuds?"

"No, he's always losing them. He screams about that, too."

"Those are good ones."

"Good what?"

"Good unsolved problems," said Debbie. "So all you have to do is try wording them with the word 'difficulty' in front. So, like, *'Difficulty keeping track of his earbuds'* would be one. And *'Difficulty keeping his music at a reasonable volume'* might be one too."

"I think I get it," said Sandra. "But he screams about a lot of things. He's going to have a ton of problems."

"Jennifer did, too. The good part is that once you identify the problems you can start solving them. That's what Kevin and I should have been doing all along."

"So, have you tried solving any problems yet?" asked Sandra.

"Not yet. We might give it a go tonight. Say, what happened at the school meeting?"

"They're not throwing him out of the program yet," said Sandra. "They just wanted to send him a message."

"A message?"

"Yeah, a message. Like he needs more messages. The only message they're sending him is that there's one more place he doesn't belong."

Here's the complete list of lagging skills that Sandra felt applied to Frankie:

> *Difficulty handling transitions, shifting from one mind-set or task to another*
> *Difficulty persisting on challenging or tedious tasks*
> *Poor sense of time*
> *Difficulty maintaining focus*
> *Difficulty considering the likely outcomes or consequences of actions (impulsive)*
> *Difficulty considering a range of solutions to a problem*
> *Difficulty expressing concerns, needs, or thoughts in words*
> *Difficulty managing emotional response to frustration so as to think rationally*

> *Chronic irritability and/or anxiety significantly impede capacity for problem solving or heighten frustration*
> *Difficulty taking into account situational factors that would suggest the need to adjust a plan of action*
> *Difficulty attending to or accurately interpreting social cues/poor perception of social nuances*
> *Difficulty appreciating how his/her behavior is affecting other people*
> *Difficulty empathizing with others, appreciating another person's perspective or point of view*

Here's a *partial* list of the unsolved problems that were associated with those lagging skills (Sandra wasn't entirely clear about the specifics of the unsolved problems that were associated with school, but she did her best):

> *Difficulty completing the homework worksheets in Science.*
> *Difficulty completing word problems on the math homework.*
> *Difficulty completing social studies homework, especially if it involves writing.*
> *Difficulty working on the project in Geography.*
> *Difficulty reading the assigned passages in Language Arts for homework.*
> *Difficulty keeping room clean.*
> *Difficulty putting clothes in laundry hamper.*
> *Difficulty waking up in the morning.*
> *Difficulty getting ready for school in time to catch the bus.*
> *Difficulty coming home on time after hanging out with his friends.*
> *Difficulty talking with Mom about problems at school.*
> *Difficulty getting along with some of the kids at school.*

> *Difficulty getting along with some of the teachers at school, especially Miss McCauley.*
> *Difficulty keeping track of his ear-buds.*
> *Difficulty keeping music at a reasonable volume when at home with Mom.*

Now let's get acquainted with a few more kids and parents you haven't yet met, but will be hearing from at various points throughout the rest of the book.

ZACH

Zach is a three-year-old preschooler who has been diagnosed with an autism spectrum disorder. He shows very little interest in interacting with other kids. At preschool, Zach seems oblivious to, and uninterested in most classroom activities. He almost always plays by himself, often speaking gibberish. Adults have to watch him closely on the playground so he doesn't wander off or stand where kids on the swing-set will crash into him. If adults try too hard to get him to participate in an activity, he screams, cries, thrashes about, and occasionally hits. At home, the television and computer are Zach's primary interests. Tantrums occur if his parents, Jonathan and Rebecca, try to separate Zach from his video activities and engage him in other activities. Left to his own devices, Zach is a very pleasant, good-natured kid.

It isn't clear whether Zach is simply uninterested in interacting with other kids or lacks the skills to do so (or both). His preferred video games are interactive, and Zach is proficient at playing them, suggesting that he *understands* language pretty well. But he uses very few words, and prefers that adults guess and figure out what he needs. He throws tantrums when adults have difficulty guessing accurately.

The new psychologist with whom Jonathan and Rebecca consulted wasn't sure about the autism spectrum diagnosis that Zach had been given. "What diagnosis we give him is the least important part," said the psychologist. "Plus, we really won't know what to call him until we figure out what's going on up there," he continued, tapping his head. The psychologist recommended that Zach's parents request an evaluation by the school system. He hoped that intensive speech and language services could be initiated as quickly as possible. In the meantime, he and Zach's parents completed the *Assessment of Lagging Skills and Unsolved Problems* together.

Zach's lagging skills include:

> Difficulty shifting from one mindset or task to another
> Difficulty expressing concerns, needs, or thoughts in words
> Difficulty managing emotional response to frustration
> Difficulty shifting from original idea, plan, or solution
> Difficulty taking into account situational factors that would suggest the need to adjust a plan of action
> Difficulty attending to or accurately interpreting social cues/poor perception of social nuances

> *Difficulty starting conversations, entering groups, connecting with people/lacking basic social skills*

Some of his unsolved problems include difficulty joining the group for circle time at school; difficulty playing with the plastic animals with another child at school; difficulty staying with the group on the school playground; difficulty joining the family for dinner; and difficulty when others don't accurately guess what he's thinking.

"So are all of these lagging skills because he's autistic?" asked Jonathan, driving the car home from the first visit with the psychologist.

"Who says he's autistic?" challenged Rebecca, who wanted to hold off on a diagnosis for as long as possible and, like the psychologist, wasn't so sure about the autism spectrum diagnosis anyway.

"So why is he lacking these skills?" asked Jonathan.

"Because he's Zach," said the mother. "Why are you so determined to label him?"

"I'm not determined to label him," Jonathan responded. "I just want to understand."

"Really, does calling him autistic help you understand him better?" said Rebecca, getting exasperated.

"Look, don't get all excited. He's my kid too."

"I'm not getting excited!" Rebecca said loudly. "I just found that whole discussion to be extremely informative. Now I finally understand what's getting in Zach's way, and I didn't know that stuff when people were telling me he was autistic."

"What if the school won't give him the help he needs unless we call him autistic?" asked Jonathan.

"Then we'll call him autistic!" Rachel exclaimed. "But we're going to get him the help he needs no matter what we call him! And the doctor is right . . . it will be a lot easier to figure out what's going on with him if we can get him talking."

MITCHELL

Mitchell is a fifteen-year-old ninth grader whose parents, Paul and Kathryn, had dragged him to yet another in a long line of therapists. At the first session, the new clinician met with Mitchell's parents before inviting Mitchell into the session. Kathryn, a social worker, and Paul, a litigation attorney, told the psychologist that Mitchell had been diagnosed with both Tourette's disorder and bipolar disorder but refused to take any medication except an antihypertensive, which reduced his tics. They also said that Mitchell was extremely unhappy about having been brought to see another mental health professional, for he hadn't found those he'd worked with previously to be especially helpful.

Kathryn and Paul reported that Mitchell, their youngest child (his siblings were already living outside the home), was extremely irritable, had no friends, and became frustrated at the drop of a hat. They indicated that Mitchell was extremely bright and very eccentric, but was repeating the ninth grade because of academic difficulties the year before.

"This is a classic case of wasted potential," Paul told the therapist. "We were devastated by what happened last year. He just plain bombed out of prep school. Here's a kid with an IQ in the 140s, and he's not making it at one of the area's top prep schools. He practically had a nervous breakdown over it. He had to be hospitalized for a week because he tried to slit his wrists."

"That sounds very serious, and very scary. How is he now?" the therapist asked.

"Lousy," said Kathryn. "He has no self-esteem left. He's lost all faith in himself. And he doesn't seem to be able to complete any schoolwork anymore. We think he's depressed."

"Where's he going to school now?" the therapist asked.

"Our local high school," Kathryn replied. "They're very nice there and everything, but we don't think he's being challenged by the work. He's so bright."

"Of course, there's more to doing well in school besides being smart," the therapist said. "Can I take a look at the testing you had done?"

Kathryn gave the therapist a copy of a psycho-educational evaluation that had been done when Mitchell was in the seventh grade. The report documented a twenty-five-point discrepancy between his exceptional verbal skills and average nonverbal skills, difficulty on tasks sensitive to distractibility, very slow processing speed, and below-average written language skills. And yet the examiner had concluded that Mitchell had no difficulties that would interfere with his learning.

"This is an interesting report," the therapist said. "It may give us some clues as to why Mitchell is struggling to live up to everyone's expectations in school."

"We were told he had no learning problems," Kathryn said.

"I think that was probably inaccurate," the therapist said. He then explained the potential ramifications of evaluation's findings. As they talked, it became clear that Mitchell was struggling the most on tasks involving a lot of writing, problem solving, rapid processing of information, and sustained effort. "That's something we're going to have to take a much closer look at," the therapist said.

"Of course, he's still very bright," said Paul.

"There are some areas in which he is clearly quite bright," the therapist said. "But there are some difficulties that may be making it very hard for him to show how bright he is. My bet is that he finds that disparity quite frustrating."

"Oh, he's frustrated, all right," said Kathryn. "We all are."

After a while, Mitchell was invited into the office. He refused to meet with the therapist alone, so his parents remained in the room.

"I'm sick of shrinks," Mitchell announced from the outset.

"How come?" the therapist asked.

"Never had much use for them. None of them has ever done me any good," Mitchell answered.

"Don't be rude, Mitchell," Paul intoned.

"SHUT UP, FATHER!" Mitchell yelled. "HE WASN'T TALKING TO YOU!"

The storm passed quickly. "It sounds like you've been through quite a bit in the past two years," the therapist said.

"WHAT DID YOU TELL HIM?!" Mitchell demanded, yelling at his parents.

"We told him about the trouble you had in school last year," Kathryn said, "and about your being suicidal, and about how we don't . . ."

"ENOUGH!" Mitchell screamed. "I just met this guy, and you've already told him my life story! I wouldn't have been suicidal if I hadn't been on about eighty-seven different medications at the time!"

"What were you taking back then?" the therapist asked.

"I don't know," Mitchell said, rubbing his forehead. "You tell him, Mother."

"I think he's been on about every psychiatric drug known to mankind," said Kathryn. "Lithium, Prozac . . ."

"STOP EXAGGERATING, MOTHER!" Mitchell boomed.

"Mitchell, don't be rude to your mother," said Paul.

"If you don't stop telling me not to be rude, I'm leaving!" Mitchell screamed.

Once again, the storm quickly subsided. "What medicines are you taking now?" the therapist asked.

"Just something for my tics," Mitchell replied, "and don't even think about telling me to take something else! I'm done talking about it!"

"He doesn't even take his tic medication all the time," said Kathryn. "That's why he still tics so much."

"MOTHER, STOP!" Mitchell boomed. "I don't care about the tics! Leave me alone about them!"

"It's just that . . ."

"MOTHER, NO!" Mitchell interrupted.

"Mitchell, are you suicidal now?" the therapist asked.

"No! And if you ask me that again, I'm leaving!"

"He still doesn't feel very good about himself, though," Paul said.

"I FEEL JUST FINE!" Mitchell yelled. "You're the ones who need a psychologist, not me!" Mitchell turned to the therapist. "Can you do something about them?"

Paul chuckled at this question.

"What's so funny?" Mitchell yelled.

"If I might interrupt," the therapist said, "I know you didn't want to be here today, and I can understand why you might not have much faith in yet another therapist. But I'm interested—what is it you'd like me to do about your parents?"

"Tell them to leave me alone," he growled. "I'm fine."

"Yes, he's got everything under complete control," Paul said sarcastically.

"PLEASE!" Mitchell said, rolling his eyes.

"If I told them to leave you alone, do you think they would?" the therapist asked.

Mitchell glared at his parents. "No, I do not."

The therapist chose his words carefully. "Is it fair to say that your interactions with your parents are very frustrating for you?"

Mitchell turned to his parents. "You've found another genius," he said. "We need to waste our time on this guy telling us the obvious?"

"Mitchell!" said Paul. "Don't be rude!"

"STOP TELLING ME WHAT TO DO!" Mitchell yelled.

"I appreciate your looking out for me," the therapist said to the father, "but I actually want to hear what Mitchell has to say." The therapist looked back at Mitchell. "I don't think I can get them to leave you alone without you being here."

"I don't think you can get them to leave me alone *with* me being here," Mitchell said. Then he paused for a moment. "How often do I have to come?" he asked.

"Well, to start, I'd like you to come every other week," the therapist said. "I'd like your parents to come every week. Is that reasonable?"

"Fine!" Mitchell said. "Can we leave now?"

Mitchell's lagging skills include difficulty managing his emotional response to frustration so as to think rationally; difficulty maintaining focus; chronic irritability and/or anxiety; difficulty handling unpredictability, ambiguity, uncertainty, novelty; difficulty appreciating how his behavior affects other people; and difficulty appreciating how he is coming across or being perceived by others. His unsolved problems include difficulty on a variety of academic tasks involving writing and sustained attention (each academic task was listed separately on the ALSUP). It may also be obvious that family communication patterns were playing a major role in fueling challenging episodes.

ANA

Ana, age seven, lives with her parents, Teresa and Ed, and her older brother and sister. She's never seen a mental health professional. She has never received a psychiatric diagnosis. But her parents are worried about her anyway. They describe her as being quite bright (they noted that she had hundreds of words in her vocabulary at 18 months, and is still quite verbal), sensitive, and intuitive beyond her years. They also describe her as intense, perfectionistic, and anxious. Initially, Ana's parents found her vocabulary and strong sense of how things "should be" to be fascinating and endearing. But over the past two years, they have become increasingly concerned about her episodes of screaming and crying when things don't go the way she expects.

When Ana was four years old, Teresa mentioned the outbursts to her pediatrician, who listened empathically but concluded that Ana was simply precocious. "Everyone's heard of the Terrible Twos," the doctor had said, "but these are the Fearsome Fours." Teresa wondered what to call it now that Ana was seven. She recently discussed the situation with Ana's maternal grandmother, Elena.

"She's a chip off the old block," laughed Elena.

"It's not funny," said Teresa. "I'm really getting worried about her. The other day she didn't like the way I was putting her clothes away in her drawers. She said there was a better way to do it. She started screaming when I tried to discuss it with her. I mean, I don't really care how things are put away in her drawers, so I just dropped the whole thing. But, geez, I can't believe I was ever that bad."

"See, that's the problem, you let her have her way," admonished Elena. "If you ever acted that way, all your father would have to do is take off his belt. You stopped."

"Look, Mom, I'm not taking a belt to my kid just because she has strong ideas," said Teresa.

"Your father didn't take a belt to you very often," said Elena. "The threat was enough. You knew who was the boss."

"I'm not threatening Ana with a belt. I'm not trying to change her personality. I want her to have opinions. I just wish she wasn't so intense about them sometimes. Plus, I send her to timeout when she's completely over the top. She knows who's the boss."

"As you wish," said Elena.

"Plus, there's a bunch of stuff a belt wouldn't fix anyway," said Teresa. "They did some program on being careful around strangers in her class at school last week and she hasn't slept in her own bed since. She thinks someone's going to break into our house and take her."

This seemed to impress Elena. "So maybe you should take her to a psychiatrist or something," she suggested.

"I looked on the Internet to see if there was a diagnosis or something that made sense," said Teresa. "Nothing."

Fortunately, the pediatrician recommended that if Teresa was truly concerned, she should take a look at the *Assessment of Lagging Skills and Unsolved Problems*. Here are some of the lagging skills she and Ed thought applied to Ana:

> *Difficulty considering a range of solutions to a problem*
> *Difficulty managing emotional response to frustration so as to think rationally*
> *Difficulty seeing the "grays"/concrete, literal, black-and-white thinking*
> *Difficulty deviating from rules, routine; difficulty handling unpredictability, ambiguity, uncertainty, novelty*
> *Difficulty shifting from original idea, plan, or solution*

And here's a partial list of the unsolved problems that were setting challenging episodes in motion:

> *Difficulty if clothes are not put away in a certain order*
> *Difficulty if her brother enters her bedroom without permission*
> *Difficulty if her brother plays with her toys*
> *Difficulty sleeping in her own bed at night*
> *Difficulty coming to dinner if her homework isn't yet completely done*

> *Difficulty completing spelling assignments if she feels her handwriting is messy*
> *Difficulty eating a wide array of foods, especially vegetables and fruits*

"So does this mean there's something wrong with her?" Ed asked Teresa.

"What do you mean, something wrong with her?"

"You know, does she have a problem?"

"Yeah, about ten of 'em," said Teresa. "Problems we could be helping her solve."

"Yeah, but how come her siblings didn't have these problems?"

Teresa mulled this question. "Actually, they did have problems, they just weren't as intense about them."

"Yeah, but look at all those lagging skills," said Ed.

"We're all lacking some skills," smiled Teresa. "I found a bunch of them that probably apply to us too."

"Should we take her to a psychologist or something?"

"I'm going to try solving some problems with her first. Then we'll see if we need a psychologist."

At this point you might be thinking, "Wow, I don't have it so bad," or, "Can we please get on with the show here? Tell me what to do!" Don't forget, we've just covered the most important thing you need to do first: figure out what skills your child is lacking and what unsolved problems are associated with those lagging skills. We'll be getting to the next step soon. For now, there's that homework assignment I told you about earlier: it's time for **you** to identify **your** child's lagging skills and unsolved problems. As you know, you can print out the *Assessment of Lagging Skills and Unsolved Problems* (ALSUP) from the *Lives in the Balance* website (again, that's www.livesinthebalance.org).

The lagging skills ensure that you have a handle on what's making life so difficult for your child. The unsolved problems help you know what to work on to reduce challenging episodes. In Chapter 6, you'll learn how to solve those problems *collaboratively* and *proactively*. The problems you solve will no longer cause challenging episodes. And when you solve problems collaboratively and proactively, you'll simultaneously teach your child many of the skills he's lacking.

One last thing. Your child is likely to have a lot of lagging skills and unsolved problems, and that can feel a little overwhelming at first. But you're not going to be able to solve all of those problems at once. There are too many of them. In fact, trying to solve them all at the same time is a very reliable way to ensure that none get solved at all. You're going to need to do some prioritizing. Any unsolved problems that are causing unsafe behavior should be a high priority. Those unsolved problems that are setting in motion challenging episodes with the greatest frequency should be high priorities as well. I usually encourage parents to pick their top three unsolved problems; those will be the ones you start trying to solve first. The rest go on the back burner for now. Some families work on only one unsolved problem in the beginning, especially if a child or the family situation is extremely unstable.

Which unsolved problems are you setting aside for now? All of the ones you haven't prioritized! But a few examples might help. Ana was very particular about what foods she was willing to eat: only certain cereals for breakfast and pizza for dinner. But Ana's parents were determined that she maintain a balanced diet, which resulted in much badgering and nagging from them (and outbursts from Ana). This unsolved problem—*Difficulty eating a wide array of foods, especially vegetables and fruits*—set in motion challenging episodes at least twice a day at breakfast and dinner. But her parents decided to set aside this particular unsolved problem in the beginning, thereby eliminating at least two challenging episodes a day and making it easier for them to focus on their initial high priorities. With Ana's help, they chose *difficulty sleeping in her own bed at night, difficulty if her brother plays with her toys,* and *difficulty if her brother enters her bedroom without permission.*

Zach routinely exhibited challenging behavior whenever his mother brought him to the supermarket, where he had difficulty with a variety of his mother's expectations, including staying next to the shopping cart, not demanding that she purchase every high-sugar cereal on the shelves, and being patient in the checkout line. But his mother, Rebecca, concluded that having Zach accompany her to the supermarket was a low priority, and she eliminated that expectation in favor of higher priorities.

OK, we're almost ready to start solving problems. We just have one more point to cover. We need to think a little more about why the things you've done already to reduce your child's outbursts may not have worked very well and may even have made things worse.

5

THE TRUTH ABOUT CONSEQUENCES

For a long time, the conventional wisdom about the cause of challenging behavior in kids has gone something like this: somewhere along the line, behaviorally challenging kids *learned* that their crying, swearing, screaming, and destructiveness brings them attention or helps them get their way by convincing their parents to give in. The corollary to this belief is that challenging behavior is planned, intentional, purposeful, and in the child's conscious control *("He's a very manipulative kid. He knows exactly what buttons to push!")*. How did the child learn these things? He learned them because his parents are passive, permissive, inconsistent disciplinarians *("What that kid needs is parents who let him know who's the boss and won't back down!"*). Parents who become convinced of this often blame themselves for their child's challenging behavior (*"We must be doing something wrong. Nothing we do seems to work with this kid."*). And if you believe that such behavior is learned and is the result of poor parenting and lax discipline, then it follows that it can also be *un*learned with better and more convincing teaching and discipline.

In general, this unlearning and re-teaching process—we'll call it the conventional intervention—includes:

1. providing the kid with lots of positive attention for good behavior and eliminating all attention associated with challenging behavior, so as to reduce the likelihood that he will seek attention by exhibiting maladaptive behavior;

2. teaching parents to issue fewer and clearer commands;

3. teaching the kid that compliance is expected and enforced on all parental commands and that he must comply quickly because his parents are going to issue a command only once or twice;

4. teaching the child that his parents won't back down in the face of challenging behavior;

5. maintaining a record-keeping and currency system (points, stickers, checks, happy faces, and the like) to track the child's performance on specified target behaviors; and

6. delivering adult-imposed consequences, in the form of rewards (such as allowance money and

privileges), loss of attention (in the form of time-
outs), and punishments (such as loss of privileges
and grounding) contingent on the child's successful or unsuccessful performance.

This conventional approach to modifying behavior isn't magic; it merely formalizes practices that have long been considered important cornerstones of effective parenting: being clear about how a child should and should not behave, consistently expecting and insisting on appropriate behavior, and giving a child the incentive to exhibit such behavior.

Some parents and their children benefit from behavior modification programs and find that they add some needed structure and organization to family disciplinary practices. However, many parents embark on such a program with an initial burst of enthusiasm, energy, and vigilance, but become less enthusiastic, energetic, and vigilant over time and return to previous patterns of parenting. And many other parents find that such programs don't improve their child's behavior at all. In fact, some parents find that such programs may actually *increase* the frequency and intensity of their kid's outbursts and cause their interactions with their kid to worsen. Let's think a little about how things could turn out that way.

Adult-imposed consequences (rewarding and punishing) basically do two things well. First, they teach kids basic lessons about right and wrong ways to behave. Of course, there are other ways besides formally rewarding and punishing to teach these basic lessons, including *direct instruction*: "Don't touch the hot stove or you'll get burned." "If you boss your friends around, they won't want to play with you." "If you don't study for your test, you won't get a good grade." Most kids learn from and adjust their behavior in response to this form of teaching. But there are some kids—the ones this book is about—who have learned the basic lessons about right and wrong but *are lacking the skills to reliably behave in a manner that is in keeping with these lessons.*

There's a second thing reward and punishment programs do well: they give kids the incentive to exhibit desirable behaviors more often and undesirable behaviors less often. But it's important to remember that adult-imposed consequences aren't the only form of consequences influencing a child's behavior. There are also those very powerful, very persuasive, inescapable, inevitable *natural* consequences. There's a powerful natural consequence for touching a hot stove, one that would teach a persuasive lesson. There are also powerful and persuasive natural consequences that will occur if a kid is too bossy with his friends or doesn't study for a test. A lot of kids learn from and adjust their behavior in response to natural consequences. Of course, there are some kids—the ones this book is about—who are already motivated to exhibit desirable behaviors more often and undesirable behaviors less often but *are lacking the skills to pull it off.*

The vast majority of behaviorally challenging kids I've worked with over the years already knew the basic lessons about right and wrong and had already endured more than their fair share of adult-imposed and natural consequences. You're safe in assuming that your child badly wants to do well, that he would do well if he could do well, that doing poorly isn't working out well for him at all, and that he's lacking the skills to do well. Maybe those rewards and punishments aren't what he needs. Maybe they don't (and weren't designed to) fix the factors that are really getting in the way. Maybe he, too, is wondering why things are still going poorly, even after all those stickers and time-outs and lost privileges. Maybe he's given up hope that doing well is in the cards. If all those consequences were going to work, they would have worked a long time ago.

Your child needs something else from you, something rewards and punishments don't deliver. If your child lacked the skills to read or spell or do math, you wouldn't use adult-imposed or natural consequences to teach those skills. Now that you've completed the ALSUP, you know what skills your child is lacking and the unsolved problems that are being caused by those lagging skills. Adult-imposed and natural consequences aren't going teach those skills or solve those problems.

Let's take a closer look at what happened back when Debbie and Kevin tried to implement a formal reward and punishment program with Jennifer. As instructed, they gave clearer directions and praised Jennifer when she behaved appropriately. They made a list of the behaviors Jennifer needed to improve on: following directions, being respectful, brushing her teeth, getting ready for bed at night, turning off her videos when instructed, and eating dinner with the family. They kept track of how Jennifer was doing on these behaviors with a point system; she received points when she met these behavioral expectations and lost points when she did not meet expectations. They made a list of tangible rewards and special privileges Jennifer would earn when she accrued a sufficient number of points. If Jennifer failed to comply with adult directives, she was given a time-out. Debbie and Kevin were now quite certain that Jennifer knew the behaviors they expected and was very motivated to perform those behaviors.

The following scenario occurred countless times. Debbie and Kevin would give a direction; for example, "Jennifer, it's time to turn off the TV and get ready for bed." In many instances, Jennifer, whose skills at shifting mind-set were not outstanding, wouldn't budge. Debbie and Kevin would repeat the directive. Jennifer would become frustrated. Her parents would then calmly remind Jennifer of the consequences of failing to comply with their expectations and directives. Jennifer, who wasn't especially enthusiastic about losing points or ending up in time-out, would become more frustrated and increasingly irrational, her control over her words and actions would diminish, and she'd begin screaming and crying. Debbie and Kevin would interpret Jennifer's increased intensity and failure to respond to their directive as an attempt to force them to "back down" or "give in" and would warn her that a time-out was imminent. Jennifer would begin throwing things at her parents. Debbie and Kevin would take Jennifer by the arm to escort her to time-out, which would further intensify her frustration and irrationality. Jennifer would resist being placed in time-out, and would try to scratch and claw her parents. They would try to restrain her physically in time-out (many clinicians no longer recommend this practice, but Jennifer's wasn't one of them); Jennifer would try to spit on or bite or head-butt them. They would confine Jennifer to her room until she calmed down. Once locked in her room—when they were actually able to get her there and keep her there—she would destroy anything she could get her hands on, including some of her favorite toys.

Ten minutes to two hours later, Jennifer's challenging episode would run its course and rationality would be restored. Debbie and Kevin would hope that what they all had just endured would eventually pay off in the form of improved compliance. When Jennifer would finally emerge from her room, she was often remorseful. Then, Debbie and Kevin would, in a firm tone, reissue the direction that started the whole episode in the first place.

Eventually, Debbie and Kevin recognized that Jennifer wasn't earning many points and was seldom earning rewards; that she was spending a great deal of time in time-out (and that they were expending a great deal of energy getting and keeping her there); and that her behavior wasn't improving at all. Indeed, the program was making them all feel worse.

If you've had a similar experience, then you're probably ready to travel down a different path. We're going to start by assuming that your child is lacking *skills* rather than *motivation*. We're going focus on *problems* rather than *behaviors*. We're going to focus on *solving those problems* rather than on *rewarding and punishing those behaviors.* And we're going to solve those problems *proactively* rather than in the heat of the moment. When the problems are solved, the challenging behaviors that are associated with those problems will subside. It won't be easy. It won't be fast. But you will start to see more and more of the better sides of your child—and of you.

QUESTION: This is all very interesting. But I'm not going to say yes to my child on everything he wants just so he doesn't get upset.
ANSWER: Good, because this book isn't going to tell you to do that.

QUESTION: Don't I need to set a precedent so my child knows who's the boss?
ANSWER: Your child already knows you're the boss. Mission accomplished. He needs you to be a different kind of boss, to use your authority in a different way.

QUESTION: So I'll still be in charge?
ANSWER: You're going to feel a lot more in charge than you do now.

We've covered a lot of territory in this chapter. Here's a summary of the key points:

- A common belief about behaviorally challenging kids is that they have learned that their challenging behavior is an effective means of getting their way and coercing adults into giving in, and that their parents are passive, permissive, inconsistent disciplinarians. If this view hasn't led to improvements in your child's behavior, you may want to try on some different lenses: your child is lacking skills rather than motivation.

- While reward and punishment programs are commonly used to modify children's behavior, there are a lot of children and families for whom these programs are ineffective. If adult-imposed and natural consequences haven't improved your child's behavior, you may want to try an alternative approach.

- Your new approach will be focused on solving problems rather than modifying behavior, centered on solving those problems collaboratively rather than through imposition of adult will, and focused less on what you do in the heat of the moment and more—much more—on what you'll be doing before problems arise.

Sandra didn't love doing laundry, but the relative solitude of the Laundromat gave her time to think. She was still mulling Frankie's latest suspension. She'd tried talking with him about it, but he'd told her to leave him alone and they'd ended up screaming at each other instead.

How did it get this bad? She thought back to how awful her life had been before she became pregnant with Frankie, and how happy she felt to have a kid of her own, even at the age of 16, a kid she was determined to treat way better than her own mom had treated her. Despite not having reliable income for a long time, they'd managed. They had fun together. Frankie was her sole focus. She took good care of him. She even avoided romantic relationships because she didn't want anything to get in the way of her raising her son. She made sure Frankie knew she expected big things from him, that she wanted him to make something of himself. He seemed really smart to her. *We were pals back then,* she recalled.

Things had been fine until Frankie started having difficulties at school in first grade. She began hearing about Frankie being hyperactive and aggressive and about the difficulties he had on various academic tasks. Frankie received extra help for his learning challenges, but for his behavioral challenges he was kept in from recess, held after school, and suspended. Various medications were tried; some made things worse and others had side effects that Frankie couldn't tolerate.

Sandra responded to Frankie's difficulties with the same determination that had seen her through other adversities, but her efforts to encourage her son to behave himself in school only led to arguments and yelling. A series of counselors guided her on using sticker charts and time-outs to deal with Frankie's challenging behaviors. He liked the idea of earning rewards in the beginning, but he became aggressive if he didn't get a reward he was aiming for and eventually lost interest. Time-out wasn't a viable option: Frankie would scream and swear and the neighbors would complain. Eventually, Frankie refused to talk with Sandra about school. If she broached the subject, he'd become violent, so violent that he'd twice been placed on inpatient psychiatry units. Having seen kids pinned to the ground and placed in seclusion rooms when they were out of control, Frankie had vowed to run away from home before he'd let himself be put in one of those places again.

Now Frankie and Sandra rarely talked to each other about much of anything.

Her thoughts turned to their new home-based mental health counselor, a guy named Matt. Their old counselor—Frankie had liked her—had been transferred to another office. They new guy wanted to put Frankie on another reward program. In their first meeting, Frankie wouldn't even look at Matt. *Can't say that I blame him,* thought Sandra, shaking her head. Matt wouldn't listen to her when she told him about how many stickers and point systems they'd been through. Sandra sighed again. *I don't know if I have the energy for this anymore.*

Sandra took a deep breath. *Enough thinking for one day.* She felt she was at a crossroads. Her son was slipping away, and the energy and determination that had sustained her through difficult times throughout her life seemed to be slipping as well. It was becoming quite clear that energy and determination—and love— weren't going to be enough to make the difference for Frankie. And it was hard to imagine that identifying lagging skills and unsolved problems was going to make much difference either.

6

THE THREE PLANS

Now we understand why and when your challenging child is challenging and why natural and adult-imposed consequences haven't made things better. You've identified your child's lagging skills and unsolved problems, and have decided on the three or four high-priority unsolved problems that you want to start working on first. You're ready to start learning about the Plans.

There are three options for dealing with those unsolved problems:

Plan A refers to solving a problem *unilaterally*, through the *imposition of adult will*.

Plan B involves solving a problem *collaboratively*.

Plan C involves *setting aside an unsolved problem*, at least for now.

If you intend to follow the guidance provided in this

book, the Plans—especially Plan B—are your future.

By the way, if your child is already meeting a given expectation, you don't need a Plan because there's no unsolved problem. For example, if your child is completing his homework to your satisfaction and without significant difficulty or conflict, you don't need a Plan because your expectation is being met. If your child is brushing his teeth to your satisfaction and without significant difficulty or conflict, you don't need a Plan because the expectation is being met. But if your child is not completing his homework or brushing his teeth in accordance with your expectations, you have an unsolved problem. You need a Plan.

Let's take a closer look at the three Plans.

PLAN A

The fact that Plan A is being described first does not mean it's the preferred Plan. As you just read, Plan A is where you're solving a problem *unilaterally*, typically through imposition of adult will. In other words, *you* are the one deciding on the solution to a given unsolved problem. Thus, the words "I've decided that . . ." are usually a good indication that you're in the midst of using Plan A: "Because you're having difficulty completing your math homework before you go outside, *I've decided that* you can't go outside until you complete your math homework," or "Because you are having such difficulty getting your teeth brushed before bed, *I've decided that* there will be no TV or video games at night until your teeth are brushed," or "*I've decided that* since you seem to be having difficulty sticking with your curfew, you're not going out with your friends at night anymore."

Now, these adult responses to unsolved problems might sound fairly ordinary, and typically they don't set the stage for challenging behavior if you have a fairly ordinary kid. But we've established that you don't have a fairly ordinary kid. With behaviorally challenging kids, Plan A actually greatly heightens the likelihood of a challenging episode. Why? For starters, your kid probably doesn't have the skills to handle Plan A. None of us is especially enthusiastic about having someone else's will imposed on us, but most of us have the skills to handle it when it happens. Behaviorally challenging kids don't. What skills? You'll find many of them listed in the lagging skills section of the *Assessment of Lagging Skills and Unsolved Problems*. Recall (from Chapter 2) that challenging episodes occur *whenever a kid doesn't have the skills to deal well with the demands that are being placed on him*. If you throw Plan A at a kid who doesn't have the skills to handle Plan A, you've placed a demand on him that outstrips his capacity to respond adaptively. That's why he's responding maladaptively. Indeed, when we "rewind the tape" on the vast majority of challenging episodes with behaviorally challenging kids, what we find is an adult using Plan A.

The paradox is that the kids least capable of handling Plan A—the behaviorally challenging ones—are the ones most likely to get it. That's because somewhere along the line (probably right around when we started viewing these kids as manipulative, attention-seeking, coercive, unmotivated, limit-testing, and so forth), many people became convinced that the best way to help behaviorally challenging kids was to apply massive doses of Plan A. Now you know that's actually the perfect recipe for massive numbers of challenging episodes.

There's another reason Plan A often isn't the best way to solve a problem. Solutions arrived at through Plan A are not only *unilateral*, they're also *uninformed*. With Plan A, you're not trying to find out why your child is having difficulty completing his math homework, or why he's having difficulty getting his teeth brushed before bed, or why he's having difficulty getting home in time for curfew. You're just insisting on having your expectations met and administering adult-imposed consequences, consequences that often inflame your child further and make it more difficult for you to gather the information you need to solve the problem. Uninformed solutions don't solve problems.

One more reason you may want to be less enthusiastic about Plan A: *being inflexible yourself isn't going to help your child be more flexible, tolerate frustration more adaptively, or solve problems more effectively*. Indeed, my experience is that being unilateral is a good way to get your kid to respond in kind. In other words, it's a good way to set the stage for frequent power struggles.

Are you dropping all the expectations you have for your child by dramatically reducing (or eliminating) your use of Plan A? No. You still have lots of expectations (and he's probably meeting many of them). Plan A is one of three options for dealing with those expectations that your child *isn't* meeting.

If Plan A isn't working for you and your kid, I'd recommend you stop using it. Of course, if you're not using Plan A, you're going to need another way to solve problems. That's Plan B.

PLAN B

Plan B involves solving a problem *collaboratively*, a process in which you and your child work together—collaborate—to solve the problems that are causing the challenging episodes that have been so destructive to your relationship with each other.

Now, according to the conventional wisdom (and many popular parenting books) you should never collaborate with a child. After all, you're in charge. But in this book, being in charge means that you understand why even the most mundane of problems can set the stage for challenging episodes, and that you're willing to change course. You're still in charge when you're using Plan B, probably more in charge than you've ever been. The only down side to Plan B is that, at least initially, it's hard to do, primarily because you may not have had much practice with it.

Plan B consists of three steps: the Empathy step, the Define the Problem step, and the Invitation step. The names of the steps aren't anywhere nearly as important as their ingredients.

1. The *Empathy step* involves gathering information from your child to understand his concern or perspective about a given unsolved problem.

2. The *Define the Problem step* involves communicating your concern or perspective about the same problem.

3. The *Invitation step* is when you and your child discuss and agree on a solution that is *realistic* (that is, you and your child can actually do what you're agreeing to do) and *mutually satisfactory* (it addresses the concerns that your child voiced in the Empathy step and that you articulated in the Define the Problem step).

These are the three ingredients that are essential to the collaborative resolution of a problem.

This next part is crucial. On first hearing about Plan B, many people come to the erroneous conclusion that the best time to use Plan B is just as they are in the midst of dealing with an unsolved problem. That's Emergency Plan B, and it's actually not the best timing because your child is already heated up (and maybe you are too). Few of us do our clearest thinking when we're heated up. And remember, the problems precipitating most challenging episodes are highly predictable. So there's no reason to wait until your child is already upset to try to solve problems that have been causing challenging episodes for a long time. The goal is to get the problem solved *ahead of time*, before it comes up again. That's Proactive Plan B.

For example, the best time to have a Plan B discussion

with your child about brushing his teeth is *long before* he's faced with the task of brushing his teeth rather than in the heat of the moment. If the unsolved problem is difficulty completing math homework, the time to have a Plan B discussion is *long before* your child is in the midst of struggling with his math homework *yet again*. Since you've already decided which high-priority unsolved problems to work on, you should be using Plan B proactively the vast majority of the time. As you'll read in the next chapter, it's often a good idea to make an appointment with your child for these problem-solving discussions.

Plan B is not a technique or a quick-fix that will magically and totally transform your child in a snap. Plan B is a process, not a recipe for overnight success. Solving problems durably, teaching skills, and changing fundamental aspects of the way you interact with your child will take time.

PLAN C

As you now know, Plan C involves temporarily setting aside an unsolved problem completely. Make no mistake: Plan C is *not* the equivalent of "giving in." Actually, giving in is what happens when you start with Plan A and end up capitulating because your child pitched a fit. The "C" of Plan C doesn't stand for "capitulating" or "caving." Plan C is about prioritizing.

Remember, you probably have a lot of problems to solve with your child, and you can't solve them all at once. You're going to focus only on your high priority unsolved problems. When you use Plan C, you intentionally and thoughtfully choose to set aside a given expectation because you have other, higher-priority expectations to pursue or because you've decided it was unrealistic in the first place. The biggest downside to Plan C is that some of your expectations won't be met right away. But the upside is that any unsolved problems you've set aside won't cause challenging episodes, so you and your kid will be better able to tackle the unsolved problems that remain.

There are a few different variants of Plan C, depending mostly on timing. If you've already identified your child's high- and low-priority unsolved problems, then you can apply Plan C *proactively* to the problems you've decided are a low priority. For example, if you've decided that brushing teeth is a low-priority unsolved problem, you simply won't tell your child to brush his teeth. If homework completion is a low-priority unsolved problem for now, then you won't tell your child to do his homework. When will you start mentioning these things again? After you've solved some higher-priority unsolved problems, you can turn your attention to those you've set aside.

Proactive Plan C can also involve coming up with an interim plan for an unsolved problem that has been set aside for the time being. Here's what that might sound like:

DEBBIE: Jennifer, you know how Dad and I are always getting on your case about eating dinner together as a family?

JENNIFER: I don't want to talk about that!

DEBBIE: Oh, I don't want to talk about it either. I just wanted to let you know that we're not going to bug you about it anymore. You don't have to eat dinner with us if you don't want to.

JENNIFER: I don't?

DEBBIE: No, there are some other problems that are more important for us to work on, so we're just going to let that one go for now.

JENNIFER: So I can eat wherever I want?

DEBBIE: Well, that's what I wanted to talk to you about for a second. I was thinking we could come up with a plan for places it's OK to eat dinner and places it's not. There are two places I'd really prefer that you not eat.

JENNIFER: Where?

DEBBIE: Your bedroom and the living room.

JENNIFER: Can I eat in the TV room?

DEBBIE: Yes, that's fine with me . . . just not your bedroom and the living room. Are you good with that?

JENNIFER: Yep. So I don't have to eat dinner with you guys if I don't want to?

DEBBIE: That's right . . . for now anyway.

JENNIFER: What if I want to eat dinner with you?

DEBBIE: You're welcome to eat dinner with us if you want to, but you don't have to. Good?

JENNIFER: Yep.

What if you slip and direct your child to do something that you've already identified as a low priority? Use Emergency Plan C and simply say OK.

PARENT: Thomas, it's time for you to brush your teeth.
THOMAS: I'm not brushing my teeth.
PARENT: OK.

In the next chapter, we're going to sink our teeth into the three ingredients of Plan B. But first, let's get some more questions answered.

Q & A

QUESTION: Let me get this straight: I'm supposed to drop all my expectations so my kid doesn't get upset anymore?
ANSWER: You're definitely not dropping all of your expectations. But it makes sense to set aside low-priority unsolved problems, because you're not going to be able to solve all the problems that are contributing to challenging episodes all at once. The unsolved problems you're setting aside for now (Plan C) will make it easier for you to work on your high-priority unsolved problems with Plan B.

QUESTION: Isn't this just about picking battles?
ANSWER: No, because it's not about battling. Battle-picking is the uncomfortable position many parents (especially those who don't know about Plan B) find themselves in: pursue an expectation through the imposition of adult will (but at the price of causing a challenging episode), or avoid a challenging episode (but at the price of not pursuing important expectations). When you add Plan B to the mix, you're not

picking battles anymore because you're not battling anymore. You're solving the problems that have been causing those battles.

QUESTION: So I'm not allowed to tell my kid what to do anymore?

ANSWER: If "telling" means you're reminding your child to fulfill an expectation he normally has no difficulty meeting, no worries. But if it's an expectation your child does have difficulty meeting—and you've already decided to set that expectation aside for now (Plan C)—then you shouldn't be telling your child to meet that expectation. And if you haven't yet identified and prioritized unsolved problems and you're simply relying on repeated "telling" to get your child to meet a given expectation (this form of "telling" is also known as "nagging"), then you may want to give serious consideration to replacing "telling" with Plan B (if it's a high priority) or Plan C (if it's not).

QUESTION: I can't set limits anymore?

ANSWER: You're setting limits when you use Plan A and you're setting limits when you use Plan B, but in completely different ways. You're probably reading this book right now because setting limits with Plan A doesn't work for you or your child.

QUESTION: What about safety issues?

ANSWER: If your child is about to step in front of a moving car in a parking lot, then of course, pull him out of the way. But if your kid *often* steps in front of moving cars in parking lots, then—unless you want to be yanking on his arm the rest of his life—you'll want to get a handle on what's causing that problem and proactively collaborate with your child to come up with a realistic and mutually satisfactory solution (Plan B).

QUESTION: So the problems I really care about, that's Plan A. And the problems I sort of care about, that's Plan B. And the problems I don't care about at all, that's Plan C. Yes?

ANSWER: No. The Plans are not a ranking system. Each Plan represents a distinct way of responding to unsolved problems. With Plan A, you're imposing a unilateral, un-informed solution and greatly heightening the likelihood of a challenging episode. With Plan C, you're temporarily setting aside the problem and reducing the likelihood of an challenging episode. With Plan B, you're identifying and clarifying concerns, working out solutions that are realistic and mutually satisfactory, and solving problems durably, so they no longer cause challenging episodes.

Here's a brief summary of what you've read in this chapter:

- There are three options for responding to unsolved problems:

  - Plan A involves solving a problem unilaterally, typically by imposing your will, and often accompanied by adult-imposed consequences (and challenging episodes).

- Plan B is where you're solving a problem collaboratively, and you're going to learn how to do that in the next chapter.

- Plan C is where you're setting aside an unsolved problem, at least for now, because there are just too many to solve all at once.

- Any unmet expectation that you can respond to with Plan A you can also respond to by using Plan B. You're setting limits with Plan A and you're setting limits with Plan B, but in very different ways. You don't lose any authority using Plan B. None.

7

PLAN B

As you read in the last chapter, Plan B has three steps: the Empathy step, the Define the Problem step, and the Invitation step. This chapter gets into the the nitty-gritty of each step, so you may find yourself referring it often. There's additional nitty-gritty—related to difficulties you may encounter in using Plan B—in Chapter 8.

THE EMPATHY STEP
The goal of the Empathy step is to *gather information from your child so you can achieve the clearest possible understanding of his concern or perspective on a given unsolved problem*. Just like adults, kids have legitimate concerns: hunger, fatigue, fear, desire (to buy or do things), the tendency to avoid things that are scary and uncomfortable or

at which they don't feel competent. You'll want to learn as much as possible about your child's concerns about a given unsolved problem.

Some adults have never thought it was especially important to gather information about and understand a kid's concern or perspective. That's why many kids—perhaps most, unfortunately—are accustomed to having their concerns ignored or dismissed by adults who have concerns of their own, or who feel that they already know what's getting in the kid's way on a given problem. Dismissing kids' concerns isn't ideal to begin with, but if you dismiss the concerns of a behaviorally challenging kid you're going to increase the likelihood of a challenging episode. Furthermore, kids who are accustomed to having their concerns dismissed tend to be far less receptive to hearing the concerns of others. If you don't understand the concerns that are fueling your kid's challenging episodes, then those concerns won't get addressed and the episodes will persist. You don't lose any authority by gathering information, understanding, and empathizing. Rather, you gain a problem-solving partner.

While many adults think they know what the child's concerns are with regard to a particular unsolved problem, when you enter the Empathy step you do so with a completely different assumption: *that you have no idea.* Well, you may have some ideas, but you're likely to find out that your ideas aren't exactly on target. So the pressure's off: there's no need to divine your child's concern or perspective. You don't need to be a mind reader. But you do need to become skilled at gathering information from your child.

The Empathy step begins with an introduction to the unsolved problem. The introduction usually begins with the words *"I've noticed that . . ."* and ends with the words *"What's up?"* In between you insert the unsolved problem. The introduction is much easier if you stick with the guidelines for writing unsolved problems delineated in Chapter 4. Here are some examples:

"I've noticed that it's been difficult for you to go to school lately. What's up?"

"I've noticed that it's been difficult for you to brush your teeth at night. What's up?"

"I've noticed that it's been difficult for you to complete your math homework lately. What's up?"

"I've noticed that it's been difficult for you to stick with the thirty-minute time limit on playing video games. What's up?"

"I've noticed that it's been hard for you to get to bed on time lately. What's up?"

"I've noticed that it's been difficult for you to get to the school bus on time lately. What's up?"

Notice that, in accordance with the guidelines, these introductions make no mention of challenging behavior or your theories about your child's concerns, that they are specific and split (not clumped); and that they include no solutions. Remember, sticking with the guidelines increases the likelihood of your child participating in the process. Since the main mission of the Empathy step is to gather information so as to understand your kid's concern or perspective about a given problem, you badly want him to talk to you. If he doesn't talk, his concern or perspective won't get addressed and the challenging behavior that occurs in association with that problem will continue.

Next comes the hardest part. After you ask "What's up?" one of five things will happen:

POSSIBILITY #1: He says something.

POSSIBILITY #2: He says nothing or "I don't know."

POSSIBILITY #3: He says, "I don't have a problem with that."

POSSIBILITY #4: He says, "I don't want to talk about it right now."

POSSIBILITY #5: He becomes defensive and says something like "I don't have to talk to you" (or worse).

Let's spend some time on each of these possibilities.

He says something.

If the introduction to an unsolved problem elicits a response, that's good. However, the initial response seldom provides a clear understanding of the child's concern or perspective, so you'll need to probe for more information. I call this probing process "drilling," and there's no doubt that drilling is the hardest of all the components of Plan B. It's where most Plan B ships run aground (and where most adults abandon ship). The good news is that there are some strategies to help you master the drilling process so the Plan B boat stays afloat.

First, notice that the word is "drill," not "grill." The primary goal of drilling is to *clarify*, whereas grilling tends to be an act of intimidation. Your goal is to demonstrate to your child that your attempt to understand his concern or perspective isn't fake or perfunctory—*you really want to understand.*

Second, *drilling* is not the same thing as *talking*. There are caregivers who frequently talk to (or at) a kid, but without trying to achieve a clear understanding of the kid's concern or perspective on a specific unsolved problem.

Why is drilling so hard? Because (1) adults think they already know what's getting in the kid's way (he's attention-seeking, manipulative, coercive, and so forth), so they don't really see the need to gather information; or (2) even when adults do want to find out, they aren't exactly sure what to say to clarify a kid's concerns. Hopefully, the early chapters of this book cast some serious doubt on #1. And, with regard to #2, those drilling strategies are going to help a lot.

Here are the drilling strategies, followed by examples:

STRATEGY #1: *Reflective listening* is the art of simply saying back to the child whatever he just said to you, and is often accompanied by clarifying statements, like "How so?" or "I don't quite understand" or "I'm confused" or "Can you say more about that?" or "What do you mean?" This is the most commonly used drilling strategy. If you're drilling and you get stuck and aren't sure what to say, reflective listening and clarifying statements are always a safe bet.

STRATEGY #2: Asking questions beginning with the words *who*, *what*, *where*, or *when*.

STRATEGY #3: Asking about why the unsolved problem occurs *under some conditions and not others*.

STRATEGY #4: Asking the child what he's *thinking* in the midst of the unsolved problem. Notice that's *thinking*, not *feeling*. While it's fine to understand a child's feelings about an unsolved problem, asking about what he's *thinking* is more likely to provide information about his concern or perspective.

STRATEGY #5: Breaking the unsolved problem down into its *component parts*. Most unsolved problems have multiple components. Getting ready for bed at night has components; so does getting ready for school in the morning and completing homework. But kids sometimes need help identifying those components so they can tell which component is causing them difficulty.

Here's an example of what drilling might sound like, with each specific drilling strategy in parentheses:

PARENT: I've noticed that we've been struggling a lot over homework lately. What's up?

ANA: It's too hard.

PARENT (USING STRATEGIES 1 AND 2): It's too hard . . . what part is too hard?

ANA: It's too much.

PARENT (STRATEGIES 1 AND 2 AGAIN): It's too much. I don't understand . . . what's too much?

ANA: The writing part is too much.

PARENT (STRATEGIES 1 AND 3): Ah, the writing part is too much. Is the writing part hard on everything?

ANA: No.

PARENT (STRATEGY 2): On what parts of your homework is the writing part too much?

ANA: I don't know.

PARENT: Well, take your time. We're not in a rush.

ANA: It's not the spelling . . . all I have to do is write one word.

PARENT (STRATEGY 1): So writing one word is not the hard part.

ANA: And it's not the social studies. All I have to do is draw a line from one word to another.

PARENT: Hmm.

ANA: It's the science part. Mrs. Moore is making us write entire paragraphs! It's too hard!

PARENT (STRATEGY 1): Ah, it's the science homework. Yes, Mrs. Moore is making you write entire paragraphs.

ANA: It's too much! It's too hard!

PARENT (STRATEGY 2): Well, I'm glad we're figuring this out. But I'm still a little confused. What is it about writing the entire paragraphs that is so hard for you?

ANA: I don't know.

PARENT (STRATEGY 5): OK, let's think about what you have to do to write the entire paragraphs. First, you have to figure out what you're supposed to write about. Is that what's hard?

ANA: No. I know what I'm supposed to write about.

PARENT (STILL STRATEGY 5): OK, then you have to figure out what you want to say in your head. Is that hard for you?

ANA: No, I know what I want to say.

PARENT (STRATEGY 5 AGAIN): OK, then you have to hold the words you want to say in your head long enough to write them down. Is that part hard?

ANA: You know I'm a slow writer! It takes me so long to write the words that I forget what I wanted to say! So then I just get all mad and then I stop doing my homework.

PARENT (STRATEGY 4): What are you thinking when that happens?

ANA: I'm thinking how stupid I am that I write so slow.

Good drilling! Very informative. We went all the way from "It's too hard" to "It takes me so long to write the words that I forget what I wanted to say" and came away with a much clearer sense of the problem that needs to be solved. If we don't have the clearest possible sense of the kid's concern or perspective, the problem won't get solved.

Adults are often astounded by what they learn when they start inquiring about a kid's concerns. Let's see what information turns up with the other examples from above (all of which would then require further drilling):

ADULT: I've noticed that it's been difficult for you to get to school lately. What's up?
KID: Sophie's been hitting me on the playground.

ADULT: I've noticed that it's been difficult for you to brush your teeth at night. What's up?
KID: I don't like the taste of the toothpaste.

ADULT: I've noticed that it's been difficult for you to stick with the thirty-minute time limit on playing video games. What's up?
KID: I don't have anyone to play with. No one in the neighborhood wants to play with me.

ADULT: I've noticed that it's been hard for you to get to bed on time lately. What's up?
KID: I don't like being alone in the dark.

ADULT: I've noticed that it's been hard for you to wake up in the morning to get to school lately. What's up?
KID: Ever since we started that new medicine, I'm really tired in the morning.

ADULT: I've noticed that it's been difficult for you to get to the school bus on time in the morning. What's up?
KID: I don't want to take the school bus anymore. The bus driver always blames me when there's trouble.

After you've made some initial headway toward understanding your kid's concerns, it can be tempting to revert to being dismissive or going back to unilateral solutions, thereby ending the conversation. Here are some examples of what *not* to do:

ADULT: I've noticed that it's been difficult for you to get to school lately. What's up?
KID: Sophie's been hitting me on the playground.
ADULT: Well, you should just hit her back.

ADULT: I've noticed that it's been difficult for you to brush your teeth at night. What's up?
KID: I don't like the taste of the toothpaste.
ADULT: I don't like the taste of the toothpaste either, but that doesn't stop me from brushing my teeth.

ADULT: I've noticed that it's been difficult for you to stick with the thirty-minute time limit on playing video games. What's up?
KID: I don't have anyone to play with. No one in the neighborhood wants to play with me.
ADULT: Oh, you have lots of friends. I think you're just making excuses.

ADULT: I've noticed that it's been hard for you to get to bed on time lately. What's up?

KID: I don't like being alone in the dark.

ADULT: Oh, you'll be fine.

ADULT: I've noticed that it's been hard for you to wake up in the morning to get to school lately. What's up?

KID: Ever since we started that new medicine, I'm really tired in the morning.

ADULT: I think you just need to try harder.

ADULT: I've noticed that it's been difficult for you to get to the school bus on time in the morning. What's up?

KID: I don't want to take the school bus anymore. The bus driver always blames me when there's trouble.

ADULT: So just stay away from the kids who cause trouble and the bus driver won't blame you.

He says nothing or "I don't know."

These responses cause panic in many caregivers. It would certainly be much easier if your kid always said something in response to "What's up?" and if he knew exactly how to explain himself. Of course, if he could, he would. Rather than giving the child time to collect his thoughts, adults often respond to silence and "I don't know" by filling the void with their own theories about the child's concern or perspective (e.g., *"I think the reason you're spending so much time playing video games is because you don't want to do your chores"*). In such instances, you've strayed quite a bit from the main goals of the Empathy step (information-gathering and understanding) and made it even more difficult for your kid to think. You may need to grow more comfortable with the silence that can occur as a kid is thinking about his concerns.

Silence and "I don't know" can mean many different things. While I discuss this in great detail in the next chapter, for now let's assume that silence and "I don't know" mean that your child hasn't given much thought to his concern or perspective on the unsolved problem you're discussing and needs some time to think about it. Your best bet is to be patient and encouraging, saying something like, *"I guess I've never asked you about this before. Take your time. We're not in a rush."*

One of the advantages of doing Plan B in a planned, proactive manner is that you're not surprising your child with the *timing* of the discussion. Indeed, it often makes good sense to make an appointment to talk and to give your child advance notice of what it is that you're going to be talking about. Otherwise, while Plan B is being done proactively, your child is still surprised by the timing and topic of the discussion, thereby increasing the likelihood of "I don't know" or silence.

He says, "I don't have a problem with that."
How can you solve a problem if your kid doesn't see it as a problem?

Well, there's actually a decent chance that there are some problems you're more concerned about than your child is (a messy room, getting to bed on time, and coming home before curfew are fairly common examples). That's not a showstopper; it's actually the beginning of learning more about his concern or perspective. The first drilling strategy (reflective listening) should serve you well in such instances: *"Ah, so you don't feel that coming home after curfew is a major concern. I don't quite understand. Can you tell me more about that?"*

Another possibility is that the kid is really saying something else (and reflective listening should again help clarify what he means). Here's an example:

PARENT: I've noticed that you've been having difficulty keeping your room clean. What's up?

KID: I don't have a problem with that.

PARENT: Ah, you don't have a problem with that. You're good with your room being messy?

KID: I didn't say that.

PARENT: I'm sorry, I thought I heard you say you didn't have a problem with your room being messy.

KID: Well, I don't have as big a problem with it as you do.

PARENT: Oh, I missed that. So you don't have as big a problem with it as I do. Do you mind your room being messy?

KID: Yeah.

PARENT: What's getting in the way of your room being cleaner?

KID: Well, at this point it's so messy, I wouldn't know where to start on cleaning it. I think I'm going to need some help.

He says, "I don't want to talk about it right now."
This response can throw adults off their game as well. A few thoughts about this scenario might help. First, he doesn't have to talk about it right now, and it's good to let him know that. Lots of kids start talking the instant they're given permission *not* to talk. Second, if he truly doesn't want to talk about it right now, he may be willing to talk about why. A lot of kids will talk about why they don't want to talk about something, which is very informative in its own right. Then, after they're through doing that, often they're comfortable enough to start talking about what they didn't want to talk about in the first place. Here's the take-away message: Don't try so hard to get the kid to talk today that you lose your credibility for tomorrow. There's always tomorrow.

He becomes defensive and says something like, "I don't have to talk to you."
Why would your kid become defensive in response to your requests for information on a particular unsolved problem? There are lots of potential reasons. Maybe he's accustomed to problems being solved unilaterally (Plan A). Maybe he thinks that if a problem is being raised he must be in trouble, so he's anticipating the lowering of the boom. Maybe he doesn't really see the point in contemplating or voicing his concerns since he's become accustomed to having them disregarded. All of these possibilities reduce the likelihood that your child will talk to you.

The best response to defensive statements is not reciprocal defensiveness but rather honesty. A good response to "*I don't have to talk to you"* would be *"You don't have to talk to me."* A good response to *"You're not my boss"* would be *"I'm not trying to boss you."* And a good response to *"You can't make me talk"* would be *"I can't make you talk."* Some reassurance that you're not using Plan A might also be helpful, as in *"I'm not telling you what to do"* (you're not), *"You're not in trouble"* (he's not), *"I'm not mad at you"* (you're not), or *"I'm just trying to understand"* (you are). Statements like *"I just want what's best for you"* and *"I'm doing this because I'm your parent and I love you"* are not helpful.

Once you have a clear understanding of your kid's concern or perspective on a given unsolved problem, you're ready to move on to the Define the Problem step. How do you know when you've reached that point? There is both a subjective answer to that question and an objective answer to that question. Here's the subjective answer: When the new information you've gathered makes sense to you. Here's the objective answer: Keep asking for more information (*"Is there anything else I need to know about this problem?"*) until your child has nothing else to say.

Are you wondering if your child has sufficient language processing and verbal communication skills to participate in Plan B? There's no question that the dialogues you've already read show what Plan B looks like with kids who have these skills. In Chapter 9 you'll read about how to adjust things for kids who are having difficulty participating in Plan B because of the lack of these skills (or for a variety of other reasons). We're just covering the basics in this chapter.

THE DEFINE THE PROBLEM STEP

The primary goal of the Define the Problem step is to enter your concern or perspective into consideration. This step usually begins with the words, *"My concern is . . ."* or *"The thing is . . ."* (you'll find many examples in the coming pages).

Like kids, adults often don't give much thought to their *concerns* about specific problems. In their eagerness to get the problem solved, adults often race past their concerns to their *solutions*. Because kids have a tendency to do the exact same thing (perhaps they've been well trained), the result is a scenario called *dueling solutions*, also known as a power struggle:

ADULT: Is your homework done?
KID: My homework's too hard.
ADULT: Your homework's too hard? It's getting late. Go do it. Now! (*adult's solution*)
KID: It's too hard! I'm not doing it! (*kid's solution*)
ADULT: Oh, you're doing it all right! Now!

Kaboom. Concerns not clarified. Problem not solved.

Fortunately, solving problems collaboratively isn't about power. Nor is it about struggling (though it can be hard). It's about clarifying the concerns of both parties and then working toward solutions that address those concerns. You're not ready to think about solutions until the concerns of both parties have been clarified. Otherwise, you won't really know what concerns the solution is intended to address.

So you'll need to give some careful thought to your concerns. If you're having difficulty figuring out what your concern is about a given problem, here's some good news: the vast majority of adult concerns are related to (1) *how the unsolved problem is affecting the kid* and/or (2) *how the unsolved problem is affecting others.*

Let's see what some typical adult concerns might be on some of the problems we considered earlier (the number at the end of each example denotes which category the concern falls into):

DIFFICULTY GOING TO SCHOOL: *The thing is, if you don't go to school, I'm concerned that you're going to miss out on a lot of important learning. Plus, we wouldn't really be solving the problem of Sophie hitting you. (1)*

DIFFICULTY BRUSHING TEETH AT NIGHT: *The thing is, if you don't brush your teeth at night, the food you've been eating all day sits on your teeth and can cause cavities. I don't want to have to spend money on the dentist. (2)*

DIFFICULTY COMPLETING MATH HOMEWORK: *My concern is that you're missing out on a lot of important practice— and getting lower grades—by not doing your homework. Plus, if you don't do the math homework, we won't know which parts of math are hard for you. (1)*

DIFFICULTY STICKING WITH THE THIRTY-MINUTE LIMIT ON VIDEO GAMES: *My concern is that all that time alone playing video games isn't making it easier for you to make friends. (1)*

DIFFICULTY GETTING TO BED ON TIME: *The thing is, when you get to bed late, you're tired at school the next day and you have trouble concentrating in your classes. (1)*

DIFFICULTY WAKING UP IN THE MORNING: *My concern is that when you have difficulty waking up in the morning, you end up being late for school, and you're falling behind in your first two classes because you're frequently not there in time to attend them. (1)*

DIFFICULTY GETTING TO THE SCHOOL BUS ON TIME: *My concern is that when you miss the school bus, I have to take you to school myself, and my boss is getting upset about me coming in late. (2)*

Let's now continue the example of Proactive Plan B we began above:

The Empathy Step

PARENT: I've noticed that we've been struggling a lot over homework lately. What's up?

ANA: It's too hard.

PARENT: It's too hard . . . what part is too hard?

ANA: It's too much.

PARENT: It's too much. I don't understand . . . what's too much?

ANA: The writing part is too much.

PARENT: Ah, the writing part is too much. Is the writing part hard on everything?

ANA: No.

PARENT: On what parts of your homework is the writing part too much?

ANA: I don't know.

PARENT: Well, take your time. We're not in a rush.

ANA: It's not the spelling . . . all I have to do is write one word.

PARENT: So writing one word is not the hard part.

ANA: And it's not the social studies. All I have to do is draw a line from one word to another.

PARENT: Hmm.

ANA: It's the science part. Mrs. Moore is making us write entire paragraphs! It's too hard!

PARENT: Ah, it's the science homework. Yes, Mrs. Moore is making you write entire paragraphs.

ANA: It's too much! It's too hard!

PARENT: Well, I'm glad we're figuring this out. But I'm still a little confused. What is it about writing the entire paragraphs that is so hard for you?

ANA: I don't know!

PARENT: OK, let's think about what you have to do to write the entire paragraphs. First, you have to figure out what you're supposed to write about. Is that what's hard?

ANA: No. I know what I'm supposed to write about.

PARENT: OK, then you have to figure out what you want to say in your head. Is that hard for you?

ANA: No, I know what I want to say.

PARENT: OK, then you have to hold the words you want to say in your head long enough to write them down. Is that part hard?

ANA: You know I'm a slow writer! It takes me so long to write the words that I forget what I wanted to say! So then I just get all mad and then I stop doing my homework.

PARENT: What are you thinking when that happens?

ANA: I'm thinking how stupid I am that I write so slow!

The Define the Problem Step

PARENT: Ah, it takes you so long to write the words and that causes you to forget what you wanted to say. That's good to know. I'm glad you told me. The thing is, if you stop doing your homework completely, then you won't get any practice at the writing and it will always be hard for you.

Two sets of concerns are on the table. There's no turning back now.

THE INVITATION STEP

This final step involves brainstorming potential solutions that address the concerns of both parties, concerns that have been identified and clarified in the first two steps. I call it the Invitation step because the adult actually invites the child to solve the problem collaboratively. The Invitation step lets the child know that solving the problem is something you're doing *with* him (collaboratively) rather than *to* him (unilaterally).

To start this step, you could simply say something like, *"Let's think about how we can solve this problem,"* or

"Let's think about how we can work that out." But to make the problem as explicit as possible, I recommend that you recap the concerns that were identified in the first two steps, usually starting with the words, *"I wonder if there's a way . . ."* In the above example, that would sound something like this: *"I wonder if there's a way for us to help you with the writing part so it doesn't take so long that you forget what you wanted to say . . ."* (the kid's concern) *" . . . but still make sure you get some practice at the writing part so it won't always be so hard for you"* (the adult's concern).

Then you give the kid the first crack at generating a solution: *"Do you have any ideas?"* This is not an indication that the burden for solving the problem is placed solely on the kid. The burden for solving the problem is placed on the Problem-Solving Partners: your child and you. But giving your kid the first crack at thinking of a solution is a good strategy because it lets him know you're actually interested in his ideas.

Many parents, in their eagerness to solve the problem, forget the Invitation step. Just as they are at the precipice of actually collaborating on a solution, they impose their will. Too often we assume that the only person capable of coming up with a good solution to a problem is the adult. While there is some chance that your kid won't be able to think of any solutions (an issue discussed in greater detail in Chapter 8), there's actually an outstanding chance your child *can* think of good solutions, ones that will take your combined concerns into account. There's also a good likelihood he has been waiting (not so patiently) for you to give him the chance.

When you use Plan B, you do so with the understanding that the solution is not predetermined. One father who had failed to remember this once said, "I don't use Plan B unless I already know how the problem is going to be solved." If you already know how the problem is going to be solved, then you're not using Plan B—you're using a "clever" form of Plan A. Plan B is not just a "clever" form of Plan A. Plan B is collaborative, Plan A is unilateral.

With Plan B, you're off the hook for coming up with instantaneous, ingenious solutions. You'd think that would be a relief for many parents, but the reality is that it takes some getting used to. Though it may have felt like coming up with a quick, unilateral solution to a problem was a time-saver, your unilateral solutions weren't working very well and were therefore taking an enormous amount of time (including the time you spend dealing with the challenging episodes that are a byproduct of those solutions). Solving a difficult problem durably requires reflection, consideration, time, a willingness to let the process of exploring solutions unfold, and, most of all, collaboration. Yes, Plan B can sometimes take a long time. But challenging episodes take much longer.

This next part is crucial (you read a little about it in the last chapter). There are two criteria for gauging whether a solution is going to get the job done, and these criteria should be considered and discussed by you and your child before signing on the dotted line: the solution must be *realistic* (meaning both parties can actually do what they're agreeing to do) and *mutually satisfactory* (meaning the solution truly and logically addresses the concerns of both parties). If a solution isn't realistic and mutually satisfactory, the problem isn't solved yet and the Problem-Solving Partners have more work to do. Your reference point for whether a solution is mutually satisfactory is whether the solution addresses the concerns of both parties. In other words, *those concerns are the reference point against which all potential solutions are weighed.*

The realistic part is important because Plan B isn't an exercise in wishful thinking. If you can't execute your part of the solution that's under consideration, don't agree to it just to end the conversation. Likewise, if you don't think your kid can execute his part of the solution that's under consideration, try to get him to take a moment to think about whether he can actually do what he's agreeing to do (*"You sure you can do that? Let's make sure we come up with a solution we can both do."*). By the way, "trying harder" is seldom a viable solution.

It is just as important that the solutions be mutually satisfactory, and that's of great comfort to adults who fear that using Plan B will cause their concerns to go unaddressed and limits to go unset. Remember, Plan A is not the only mechanism by which adults can set limits; Plan B gets the job done, too. If a solution is mutually satisfactory, then by definition your concerns have been addressed. And if your concerns have been addressed, then you're setting limits.

Coming up with mutually satisfactory solutions can be hard. Early on, many kids have a tendency to think of solutions that will address their own concerns but not yours (many adults have the same tendency). But if you want him to be thinking rather than getting upset, the last thing you'd want to do is tell him he's come up with a bad idea. Instead, simply remind him that the goal is to come up with a solution that works for both of you, perhaps by saying, *"Well, that's an idea, and I know that idea would address your concern, but I don't think it would address my concern. Let's see if we can come up with an idea that will work for both of us."* In other words, there's no such thing as a bad solution—only solutions that aren't realistic or mutually satisfactory.

You want your kid to learn that *you're as invested in ensuring that his concerns are addressed as you are in making sure that yours are addressed.* That's how you lose an enemy and gain a problem-solving partner. That's how you move from adversary to teammate. This is how relationships are rebuilt and communication restored. This is also how both parties (kids *and* adults) learn new skills (this is discussed further in Chapter 9). Kids whose concerns are being heard, clarified, understood, validated, and addressed rather than dismissed or ignored become more interested in hearing your concerns and making sure that they are addressed too.

Let's see how the three ingredients go together, assuming that things are going smoothly. Forgive the redundancy, but it's nice to see the process unfold from start to finish.

The Empathy Step

PARENT: I've noticed that we've been struggling a lot over homework lately. What's up?

ANA: It's too hard.

PARENT: It's too hard . . . which part is too hard?

ANA: It's too much.

PARENT: It's too much. I don't understand . . . what's too much?

ANA: The writing part is too much.

PARENT: Ah, the writing part is too much. Is the writing part hard on everything?

ANA: No.

PARENT: On what parts of your homework is the writing part too much?

ANA: I don't know.

PARENT: Well, take your time. We're not in a rush.

ANA: It's not the spelling . . . all I have to do is write one word.

PARENT: So writing one word is not the hard part.

ANA: And it's not the social studies. All I have to do is draw a line from one word to another.

PARENT: Hmm.

ANA: It's the science part. Mrs. Moore is making us write entire paragraphs! It's too hard!

PARENT: Ah, it's the science homework. Yes, Mrs. Moore is making you write entire paragraphs.

ANA: It's too much! It's too hard!

PARENT: Well, I'm glad we're figuring this out. But I'm still a little confused. What is it about writing the entire paragraphs that is so hard for you?

ANA: I don't know!

PARENT: OK . . . let's think about what you have to do to write the entire paragraphs. First, you have to figure out what you're supposed to write about. Is that what's hard?

ANA: No. I know what I'm supposed to write about.

PARENT: OK, then you have to figure out what you want to say in your head. Is that hard for you?

ANA: No, I know what I want to say.

PARENT: OK, then you have to hold the words you want to say in your head long enough to write them down. Is that part hard?

ANA: You know I'm a slow writer! It takes me so long to write the words that I forget what I wanted to say! So then I just get all mad and then I stop doing my homework.

PARENT: What are you thinking when that happens?

ANA: I'm thinking how stupid I am that I write so slow!

The Define the Problem Step

PARENT: Ah, it takes you so long to write the words, and that causes you to forget what you wanted to say. That's good to know. I'm glad you told me. The thing is, if you stop doing your homework completely, then you won't get any practice at writing and it will always be hard for you.

The Invitation Step

PARENT: I wonder if there's a way for us to help you with the writing part so it doesn't take so long that you forget what you wanted to say, but also make sure you get some practice at the writing part so it won't always be so hard for you. Do you have any ideas?

ANA: Um . . . no.

PARENT: Well, take your time. We've never really talked about it like this before. If you don't have any ideas, maybe I can come up with some.

ANA: If we could just come up with a way for me to remember what I wanted to write. Then maybe I wouldn't get so frustrated.

PARENT: Let's think about that. How could we help you remember what you wanted to write?

ANA: At school, sometimes they have someone scribe for me.

PARENT: I know, but the last time I asked if I could scribe for you, they said they wanted you to practice your writing on your homework. Of course, they don't know what we go through around here during homework. I could ask them again, but I wonder if there's any other way we could help you remember what you wanted to write.

ANA: I could say what I wanted to write into your voice recorder. You know, how you do for work. Then I could just play the recording back and write it down.

PARENT: That could work. Would that make it easier for you to remember what you want to write?

ANA: I think so.

PARENT: Well, it sounds like that solution works for you, and it certainly works for me. I don't use my voice recorder when you're doing your homework, so it sounds realistic. Shall we give it a try?

ANA: OK.

PARENT: And if that solution doesn't work, we'll talk some more and come up with one that does.
ANA: OK.

The last thing Ana's parent said to her was key, as it underscores a very important point: it's good for both the child and the adult to acknowledge that the problem may require additional discussion, because there's a good chance that *the first solution won't solve the problem durably*. Why not? Often because the solution isn't as realistic or mutually satisfactory as it first seems. Or because the first attempt at clarifying concerns yielded useful but incomplete information. By definition, the solution will only address the concerns you know about, but can't possibly address the ones you haven't heard about yet. Solving a problem that has been causing major disagreements for a long time often isn't a one-shot deal. Good solutions—durable ones—are usually refined versions of the solutions that came before them.

Plan B isn't usually as seamless as I've depicted in the above dialogue, especially early on. Sometimes kids (and even adults) get pretty heated up while using Plan B. Sometimes this is because, historically, disagreements have been handled using Plan A. It may take a while (and a lot of Plan B) for the child's instantaneous heated reaction to unsolved problems to subside. Adults sometimes become impatient in the midst of Plan B and head for Plan A or Plan C. Hang in there.

So far, I've only provided examples of Proactive Plan B. Are you wondering what Emergency Plan B would sound like? Here are a few examples, though I can't resist reminding you that Proactive Plan B is far preferable and that if you've used the ALSUP to identify your child's unsolved problems, the vast majority of problem solving you should be doing is of the proactive variety. That said, the Empathy step of Emergency Plan B wouldn't begin with an Introduction (as in Proactive Plan B) because it's already too late. So you'd head straight into reflective listening. Here are a few examples of what that might sound like:

KID: I'm not taking my meds.
ADULT: You're not taking your meds. What's up?

KID: I'm not going to school today.
ADULT: You're not going to school today. What's up?

KID: This homework sucks!
ADULT: You're getting frustrated about your homework. What's up?

The Define the Problem step and the Invitation step are much the same with Emergency Plan B as with Proactive Plan B (though things are often louder and more intense under emergent conditions). Because Emergency Plan B typically occurs in hurried conditions and after a kid is already heated up, it isn't ideal for gathering information and solving problems durably. So, while Emergency Plan B is available to you as an option, you don't want to make a habit of it.

Now, a caveat: Proactive Plan B is generally far preferable to Emergency Plan B, but there are some kids who have difficulty with Proactive Plan B because they have trouble remembering the specifics of the problems you're trying to discuss. For these kids, the problem is only memorable and salient when they're in the midst of it. Early on, Emergency Plan B may actually be preferable for these kids. I've found that many of these kids are able to participate in proactive discussions once Plan B becomes more familiar to them.

In Chapter 3, you were given your first homework assignment: use the ALSUP to identify your child's lagging skills and unsolved problems and then identify your top-priority unsolved problems. Here's your next assignment, and it's going to be harder than the first: pick one of your high-priority unsolved problems, make an appointment with your kid, and try using Proactive Plan B to solve it. If it goes well, terrific. If it doesn't go well—and, this being a new skill, there's a decent chance it won't—keep reading.

Here's a brief summary of what you've just read:

- Plan B consists of three steps or ingredients:

    1. **The Empathy Step:** Gathering information and understanding your child's concern about a given problem.
    2. **The Define the Problem Step:** Being specific about your concern or perspective about the same problem.
    3. **The Invitation:** Brainstorming with your child to find solutions that are realistic and mutually satisfactory.

- There are two forms of Plan B, depending on timing: Emergency Plan B and Proactive Plan B. Because Proactive Plan B is far preferable, it's been the primary focus of this chapter. Emergency Plan B—because of added heat and time pressure—is much harder, and much less likely to lead to durable solutions. • With Proactive Plan B, the Empathy step begins with an introduction (*"I've noticed that..."*) to one of your high-priority unsolved problems, followed by an inquiry (*What's up?"*). The first thing your child says (if he says anything) isn't likely to provide sufficient information, so you'll want to probe (drill) for more information. Keep drilling until you feel you have a clear understanding of your kid's concern about or perspective on the problem. • The Define the Problem step usually begins with the words *"My concern is . . ."* or *"The thing is . . ."* You may want to give some thought to your concern about or perspective on the problem ahead of time. • In the Invitation (*"I wonder if there's a way..."*) you summarize the concerns that have been clarified in the first two steps and then give the kid the first crack at coming up with solutions. You aren't ready to begin thinking about solutions until you've identified and clarified the concerns of both parties.

- Like any new skill, Plan B is hard, and it takes time to feel comfortable with it. The more you practice, the easier Plan B becomes. Plan B isn't something you do two or three times before returning to your old way of doing things. It's not a technique; it's a way of life.

Debbie and Kevin had agreed that it might be less overwhelming for everyone if Debbie tried Plan B on her own with Jennifer the first time. A few days earlier, Debbie had asked Jennifer if it might be possible for them to talk about something over the weekend. Debbie thought Jennifer would balk at the idea, so she was surprised when Jennifer agreed. While Debbie knew it would be best to give Jennifer advance notice of what she wanted to talk about, Jennifer didn't ask and Debbie feared that Jennifer would refuse to talk if she was explicit about the topic. They agreed to talk during breakfast on Saturday morning. Kevin and Riley were at hockey practice.

"Jennifer, remember we agreed to talk about something during breakfast this morning?" Debbie began, sitting down at the kitchen table with Jennifer.

Jennifer grunted through a mouthful of waffles.

Debbie continued. "I was hoping we could talk about the difficulty you and Riley have when you're watching TV together."

"He should just let me watch what I want. I'm the older sister," said Jennifer.

Debbie knew Jennifer had just proposed a solution and that they weren't supposed to be talking about solutions yet. "That's an interesting idea." Debbie wasn't sure what to say next. Then she remembered that her default strategy was reflective listening. "So you're the older sister and you feel you should be able to watch what you want."

"Uh-huh."

Debbie was briefly at a loss. She was pretty surprised that Jennifer was participating in the conversation and not screaming or running out of the room. That was good. But what to say next? Debbie opted for a clarifying question. "Can you tell me more about that?"

Jennifer wiped some maple syrup from her lips. "Not really."

This is hard! thought Debbie. She tried to remember the drilling strategies and resisted the temptation to jump to the Define the Problem step. Though the silence felt interminable to Debbie, Jennifer didn't seem to mind. Debbie went with a different drilling strategy. "You know, I'm not even sure I know what it is that you guys are disagreeing about when you're watching TV. Can you tell me about that?"

"Riley always wants to watch *SportsCenter*, and I hate *SportsCenter*. All he thinks about is sports."

More info! thought Debbie. She stuck with reflective listening. "So Riley always wants to watch *SportsCenter*." Then she shifted to the second drilling strategy. "What do you want to watch?"

"Anything besides *SportsCenter*," said Jennifer. "I like *Dance Moms*. Or *Say Yes to the Dress*. He hates those shows." Jennifer paused. "Why are we talking about this anyway? He should just let me watch what I want. I'm the older sister."

Debbie was briefly stumped by Jennifer's return to her previous stance, but was curious about her daughter's perspective. "Tell me more about that."

"I'm the older sister."

"Yes, you are the older sister. But help me understand why you should pick what's on the TV."

"Because I was here first."

Debbie was alarmed to see that Jennifer was now getting up from the table. "Where are you going, honey?"

"I'm done with my breakfast," said Jennifer.

"Yes, but we're not done talking," said Debbie.

"I am," said Jennifer. She left the kitchen and went to her room.

Debbie hadn't expected the conversation to end so abruptly,

though it had lasted much longer than she'd anticipated. She tried to process what had just happened. On the one hand, she was sorry that the conversation hadn't lasted longer. They didn't even make it all the way through the Empathy step. On the other hand, Jennifer talked! She provided some information! She didn't blow up! Maybe she'll talk again . . .

"Problem solving is incremental," Debbie whispered to herself, quoting something she'd read on the new website.

8

TROUBLE IN PARADISE

In Chapter 6 you learned about the three options for responding to unsolved problems, and in Chapter 7 you learned quite a bit about Plan B. You were also given your second homework assignment: try to solve a problem using Proactive Plan B.

How did your first attempt at Plan B go? If in the Empathy step you learned about your child's concerns on a given unsolved problem, that's fantastic. If in the Define the Problem step you resisted the temptation to put your solutions on the table and instead were able to identify your own concerns, that's outstanding. If you made it to the Invitation and were able to collaborate with your child on a realistic and mutually satisfactory solution, that's excellent. Hopefully, the solution you and your child agreed upon will stand the test of time. If it doesn't, you'll find out soon enough, and then it's back to Plan B to figure out why and to come up with a solution that is more realistic or mutually satisfactory than the first one, or one that addresses concerns that may not have been identified in your first try. When you think the time is right, move on to another unsolved problem.

If things didn't go so well, don't despair. As you already know, it can take a while for you and your child to become good at this. Plan B can go astray for a variety of reasons. Let's take a closer look at some patterns that may be getting in the way.

You haven't tried Plan B yet. You're still only using Plan A and Plan C.

Maybe you don't feel very confident about your Plan B skills, so you're a little reluctant to give it a whirl. That's understandable, given that you've probably had a lot more practice at Plan A. Or perhaps you're worried that your child will respond to Plan B with the same heated and volatile reaction that he always has to Plan A. We can't rule out that possibility completely; some kids are so accustomed to Plan A that they don't immediately recognize that you're trying hard to solve problems in a different way. So you may have some residual heat to contend with. On the other hand, the information gathering and understanding that occurs in the Empathy step are very powerful ingredients, especially if you're using Plan B proactively rather than emergently. It comes down to this: if you never give Plan B a try, then you and your kid will never become good at it. No one is great at Plan B in the beginning. You and your child are learning to do this together.

You've tried Plan B, but you're relying primarily on Emergency Plan B instead of Proactive Plan B.

Remember, Emergency Plan B involves more heat (as in heat of the moment), more rush (as in you're in the middle of something or on your way somewhere), and less ideal circumstances (for example, you're driving the car, in a parking lot, or in the middle of a department store, and have other kids and people around). All of those factors work against you when you're trying to solve a problem collaboratively. You have much better odds when you going about it proactively. That's why the *Assessment of Lagging Skills and Unsolved Problems* is so important: it sets the stage for you to identify unsolved problems and decide on high-priority unsolved problems ahead of time.

If you're not especially methodical and organized, being proactive can be a challenge. But being in perpetual crisis mode is even more challenging. Solving problems collaboratively, improving your relationship with your kid, and helping him learn the skills he needs to be more flexible and handle frustration more adaptively isn't going to be easy, and is likely to require that you make some adjustments to your standard operating procedure.

If you're extremely busy and are accustomed to solving problems in the heat of the moment, there's a good chance you're leaving your kid floundering in your wake, and he's probably not faring so well back there. You could demand that *he* adapt to *you*, but since flexibility and adaptability are not his strengths, the more realistic option is for *you* to adapt to *him*. Once he learns some skills and you are able to solve some chronic problems together, he may be able to reciprocate.

You're using Plan B as a last resort.

Plan B isn't an act of desperation, and it's not something you turn to only when you're on the verge of a challenging episode (i.e., after you've just used Plan A).

You still have your old lenses on.

If you're still not convinced that your kid lacks the skills to be flexible, handle frustration, and solve problems, you may want to reread Chapters 2 and 3. Don't forget, the alternative explanation—that your child is attention-seeking, manipulative, coercive, limit-testing, and unmotivated, and that you're a passive, permissive, inconsistent, non-contingent disciplinarian—hasn't made things any better, so you really don't have a lot to lose by trying on different lenses.

You're beginning the Empathy step thinking you already know your child's concern or perspective.

As you read earlier, it's common for adults to make incorrect assumptions about a child's concern or perspective on an unsolved problem. If you enter the Empathy step quite certain that you already know his concern, you're at risk for perfunctory drilling and/or for steering the ship toward a predetermined dock. But you still won't have the information you need to solve the problem.

You're entering Plan B with a preordained solution.
It's fine to have some ideas for how a problem can be solved, but it's impossible to know what the solution is until you've identified the concerns of both parties. Remember, the reference point for all solutions is the degree to which it addresses the concerns of both you and your child.

You're agreeing to solutions that aren't realistic and mutually satisfactory.
Before you sign off on a solution, make sure you and your child have considered whether the solution under consideration is truly realistic (meaning both parties can reliably follow through on what they're agreeing to do) and mutually satisfactory (meaning the concerns of both parties have truly and logically been addressed). If there's any doubt about whether a solution is realistic and mutually satisfactory, continue discussing alternatives until you and your child agree on a solution that comes closer to the mark.

You're trying to bake the cake without one of the three key ingredients.
Each of the three ingredients, each step, is indispensable in the collaborative resolution of a problem. If you skip the Empathy step, whatever solution you come up with will be uninformed and address only your concerns. Those solutions don't work very well. Case in point:

> **ADULT:** I want to make sure you get your homework done before soccer practice from now on because if you don't

do your homework before soccer practice, it doesn't get done. How can we work that out?

KID: Huh?

As you now know, the Define the Problem step involves entering your concern or perspective into consideration. But many adults enter a solution rather than a concern in this step, causing Plan B to morph into Plan A. Let's see what that looks like:

ADULT: I've noticed that it's difficult for you to complete your homework on days that you have soccer practice. What's up?

KID: Well, I'm really tired when I get home from soccer practice, and by the time we get through with dinner it's really late.

ADULT (USING REFLECTIVE LISTENING): So you're really tired when you get home from soccer practice and it's really late after we finish dinner.

KID: Yeah, and I always mean to get up early the next morning and do the homework, but then I'm really tired in the morning too.

ADULT (USING MORE REFLECTIVE LISTENING): Ah, so you always think you'll get up to do it in the morning, but you're too tired in the morning too.

KID: Yeah.

ADULT (CHECKING IN TO FIND OUT IF THERE'S ANYTHING MORE): Anything else I should know about why it's hard for you to do your homework on the days you have soccer practice?

KID: No, that's it.

ADULT (VOICING A SOLUTION RATHER THAN A CONCERN): Well, my concern is that if you're too tired after soccer practice and you're too tired to do the homework the next morning, you need to do the homework before soccer practice.

KID: I don't want to do it before soccer practice! I'm tired when I get home from school and I need some time to chill!

Many adults get through the first two steps of Plan B, but then skip the Invitation step and impose a solution. Sometimes this is because the adults still can't fathom that a child might be able to collaborate on a realistic and mutually satisfactory solution. Most often, it's just a bad habit.

ADULT: I've noticed that it's difficult for you to complete your homework on days that you have soccer practice. What's up?

KID: Well, I'm really tired when I get home from soccer practice, and then by the time we get through with dinner, it's really late.

ADULT: So you're really tired when get you home from soccer practice and it's really late after we finish dinner.

KID: Yeah, and I always mean to get up early the next morning and do the homework, but then I'm really tired in the morning too.

ADULT: Ah, so you always think you'll get up to do it in the morning, but you're too tired in the morning, too.

KID: Yeah.

ADULT: Anything else I should know about why it's hard for you to do your homework on the days you have soccer practice?

KID: No, that's it.

ADULT (THIS TIME VOICING A CONCERN IN THE DEFINE THE PROBLEM STEP): I think I understand. My concern is that if you're too tired to do your homework after soccer, and you're too tired to do it early the next morning, then it ends up not getting done, and that's starting to affect your grades in your classes.

KID: I know.

ADULT (SKIPPING THE INVITATION STEP AND HEADING STRAIGHT INTO A UNILATERAL SOLUTION): So I've decided that if the homework isn't completely done before soccer practice then you can't go to soccer practice.

KID: What?!

ADULT (USING ONE OF THE CLASSIC RATIONALES FOR PLAN A): I'm doing this for your own good.

KID: Well, that's a crappy idea and I'm not doing it!

ADULT: Watch your tone, young man . . .

The Empathy step never gets rolling because your kid's first response to the unsolved problem is "I don't know" or silence.

As you read in the last chapter, "I don't know" or silence causes many people to get stuck in the Plan B mud. Remember, your best initial strategy is to give the child some time to think. Writing unsolved problems according to the guidelines can also reduce the likelihood of "I don't know" and silence, so you may want to double check your wording. Using Plan B proactively so your child isn't surprised by your desire to have a discussion and giving him some advance notice about the topic can reduce the likelihood of "I don't know" and silence as well. But if you're in good shape on all those counts and you still find yourself dealing with "I don't know" or silence, you'll need to figure out what the "I don't know" or silence means. Here's the short list of possibilities:

IT'S POSSIBLE HE REALLY DOESN'T KNOW WHAT HIS CONCERN IS. Perhaps you've never inquired about his concerns before, at least not in this way. Perhaps he's never given the matter any thought. Perhaps he's become so accustomed to having his concerns dismissed that he hasn't thought about his concerns for a very long time. Proactive Plan B will help him give the matter some thought, as long as you're not talking while he's trying to think. A lot of adults aren't comfortable with the silence that can occur as a kid is thinking about his concerns. Remember, if you're talking while your kid is trying to think, you'll make it harder for him, thereby reducing your chances of gathering information about his concern as well as the likelihood that his concern will be addressed. He may also need some reassurance that you're not mad, that he's not in trouble, that you're not going to tell him what to do, and that you truly just want to understand.

HE'S HAD SO MUCH PLAN A IN HIS LIFE THAT HE'S STILL BETTING ON THE PLAN A HORSE. You'll have to reassure him that you're not riding that horse anymore. By the way, mere reassurance about that may not be enough. The proof's in the pudding.

HE THINKS HE'S IN TROUBLE. History has taught a lot of kids that discussing a problem means they're in hot water and that adult-imposed consequences are forthcoming. You'll have to prove otherwise. If your child misperceives Plan B discussions as a sign of hot water, a discussion about that may be necessary, so you can learn as much as possible about the ways in which "problem solving" still feels like "trouble" to your child and come up with a solution so it doesn't feel that way anymore.

HE MAY HAVE SOME THINGS TO SAY THAT HE KNOWS YOU DON'T WANT TO HEAR, AND HE THINKS IF HE SAYS THOSE THINGS, IT'LL CAUSE A FIGHT. Your goal in the Empathy step is to suspend your emotional response to what your child is saying, knowing that if you react emotionally to what you're hearing you won't end up hearing anything. Remember, you badly want to know your child's concerns. If you don't know what his concerns are, those concerns won't get addressed and the problem will remain unsolved.

HE FORGOT OR DIDN'T UNDERSTAND WHAT YOU ASKED. If he doesn't verbalize this, his facial expression will likely provide some hints. You can always ask, *"Do you remember my question?"* or *"Do you understand what I'm asking?"* Repeat or clarify the question if necessary.

HE'S HAVING TROUBLE PUTTING HIS THOUGHTS INTO WORDS.
Some clarification might help: *"Do you know what you
want to say but you're having trouble finding the words to
say it? Or do you not know what you want to say?"*

HE'S BUYING TIME. A lot of kids say *"I don't know"* instead of *"umm,"* or *"give me a second,"* or *"let me
think about that a minute."* Since you're not in a rush,
you'll be able to give him a second and let him think
about it a minute.

If, after you've given your kid the chance to think,
you become convinced that he really has no idea what his
concern is, or that he is unable to put his thoughts into
words, your best option is to do some educated guessing
or hypothesis testing. Suggest a few possibilities, based
on experience, and see if any resonate. The good news is
that for each problem there are a finite number of potential concerns. For example, while it might feel as though
there is a universe of possible concerns related to your
child having difficulty sticking to the allotted amount
of screen time, there are probably only four or five. And
there are probably only four or five possible concerns for
each of the other problems you and your kid are trying
to solve together, too. Here's an example of educated
guessing:

ADULT: I've noticed that you haven't been too enthusiastic about taking your medicine. What's up?
KID: I don't know.

ADULT: Well, let's think about it. There's no rush.

KID (AFTER TEN SECONDS): I really don't know.

ADULT: Take your time. Let's see if we can figure it out.

KID (AFTER ANOTHER FIVE SECONDS): I really don't know.

ADULT: OK. You know we've run into this problem a few times before. Should we think about what it's been before?

KID: I can't remember.

ADULT: Well, sometimes it looks like you're having trouble swallowing the pill. Is that it?

KID: No.

ADULT: Sometimes it makes you sick to your stomach. Is that the problem now?

KID: Um, no.

ADULT: Does it bother you that you have to take it at school and the other kids see you going down to the nurse?

KID: Yes!

ADULT: Ah, so that's it. Anything else that we're not thinking of?

KID: I don't think so.

Good, now we're in the ballpark, so we can drill further. As you're in the midst of hypothesizing, bear in mind that you're proposing possibilities rather divining the kid's concern. Here's what divining sounds like:

ADULT: I've noticed that you haven't been too enthusiastic about taking your medicine. What's up?

KID: I don't know.

ADULT: I think it's because you're having trouble swallowing the pill. I thought we were over that, but I guess not.

You got stuck in the Empathy step because you had trouble drilling. It's not always easy to know what to say to keep your kid talking so you can get the information you're seeking, and there are some things kids say in response to "What's up?" that can be especially vexing. Here are some examples:

ADULT: I've noticed we've been struggling a lot on your homework lately. What's up?
KID: It's boring.
ADULT (TRYING TO DRILL A LITTLE): What's boring about it?
KID: It's just boring.
ADULT (STILL TRYING TO DRILL): Well, can you tell me some of the assignments that you're finding boring?
KID: My mind is a complete blank.

ADULT: I've noticed you haven't been eating what I make for dinner lately. What's up?
KID: I don't like it.
ADULT (TRYING TO DRILL): What don't you like about it?
KID: It doesn't taste good.
ADULT (STILL TRYING TO DRILL): Well, can you tell me what doesn't taste good?
KID: It just doesn't taste good.

When initial attempts at drilling don't strike oil, you may be inclined to abandon the well. Hang in there. You always have educated guessing or hypothesis testing as a last resort. But again, your best default drilling option is reflective listening: simply saying back to the child whatever he just said, accompanied by a clarifying statement. Let's see what this drilling strategy (and others) might look like in situations in which it appears the well is dry. These dialogues don't take you all the way through Plan B; they focus solely on drilling perseverance. (The drilling strategy appears in parentheses.)

ADULT: I've noticed we've been struggling a lot on your homework lately. What's up?

KID: It's boring.

ADULT (ASKING A QUESTION BEGINNING WITH WHO, WHAT, WHERE, OR WHEN): What's boring about it?

KID: It's just boring.

ADULT (ASKING A QUESTION BEGINNING WITH WHO, WHAT, WHERE, OR WHEN): What assignments are boring?

KID: My mind is a complete blank.

ADULT (NOT ABANDONING THE WELL AND ASKING THE KID WHAT HE'S THINKING IN THE MIDST OF THE UNSOLVED PROBLEM): Hmm. So when you're sitting there trying to do your homework, what are you thinking?

KID: I'm thinking it's boring.

ADULT (USING REFLECTIVE LISTENING AND ASKING FOR MORE OF WHAT THE KID IS THINKING IN THE MIDST OF THE UNSOLVED PROBLEM): Ah, you're thinking it's boring. What else are you thinking?

KID: I'm thinking I don't understand it.

ADULT: Is there a certain part you don't understand?

KID: The math. I just don't get it.

ADULT (BREAKING THE UNSOLVED PROBLEM DOWN INTO ITS COMPONENT PARTS): OK, let's think about the parts of the math that you're not understanding . . .

Of course, the conversation would continue from there. Way to hang in there!

Let's try another:

ADULT: I've noticed you haven't been eating what I make for dinner lately. What's up?

KID: I don't like it.

ADULT (ASKING A QUESTION BEGINNING WITH WHO, WHAT, WHERE, OR WHEN): What don't you like about it?

KID: I just don't like it.

ADULT (REFLECTIVE LISTENING): You just don't like it. Can you say more about that?

KID: It just doesn't taste good.

ADULT (REFLECTIVE LISTENING): Ah, it just doesn't taste good. What do you mean?

KID: I don't know.

ADULT (NOT ABANDONING THE WELL AND ASKING ABOUT WHY THE PROBLEM OCCURS UNDER SOME CONDITIONS AND NOT OTHERS): You know, I've noticed that some nights you eat what I make and some nights you don't. Are there some things I make that you like and some things I make that you don't?

KID: I like pasta.

ADULT: Yes, I've noticed that you do like pasta. But I think there are other things I make that you eat.

KID: Like what?

ADULT: Rice.

KID: Oh, yeah, rice. But when you put all that stuff in it, like nuts, and those little slices of orange, it's disgusting.

ADULT: Anything else I make that you like?

KID: No.

ADULT: Anything I make that you especially don't like? I mean, besides the rice with the nuts and mandarin oranges in it.

KID: Well, I kinda like your meatballs, but that's it. And I don't like the vegetables . . . except corn on the cob.

ADULT: I'm glad we're figuring out what you like and don't like. That'll help us solve this problem.

Your kid verbalized his concern or perspective in the Empathy step, but you didn't believe him.

While it's conceivable that your kid's first stab at identifying and articulating his concerns may not be spot on (after all, he may not have given his concerns much thought until you asked), a lot of adults are quick to view a kid's concerns as wrong or untrue. But the last thing you'd want to do is dismiss his concern, or worse, tell him you think he's lying. That approach only increases the likelihood that he'll stop talking to you. I've found that most of the concerns that adults thought were wrong or untrue actually had a kernel of truth to them. If you're drilling well, you'll give your kid the opportunity to clarify his concerns.

When adults tell me they think a kid is lying in the Empathy step, it's often because the adult isn't inquiring about a specific unsolved problem but rather about a *behavior* someone saw the kid exhibit. This often causes the adult to *grill* rather than drill. Here's what that sounds like (notice the adult isn't really doing the Empathy step):

PARENT: I heard from your teacher, Ms. Adams, that you hit Jovan on the playground.

KID: I did not. She's lying.

PARENT: Now why would Ms. Adams lie about that?

KID: I don't know, but she is. I didn't hit him. He hit me.

PARENT: That's not what she said.

KID: Well she's wrong.

PARENT: She said she saw it with her own eyes!

KID: Then she's blind, 'cause I didn't hit him. He hit me. Why don't you believe me?

Whether the kid is telling the truth or not is one issue (we all know how unreliable eyewitness accounts can be). But trying to get to the bottom of a specific incident is beside the point anyway, because what happened during a specific incident isn't nearly as important as solving the chronic problem: the kid and Jovan are having difficulty getting along on the playground.

Your kid said he didn't care about your concern, so your enthusiasm for Plan B dissipated rapidly.

Don't be insulted that he doesn't care about your concern. The good news is that he doesn't really have to *care* about your concern; he just has to take it into account as you pursue a mutually satisfactory solution together. He'll start trying to address your concerns not too long after you start trying to address his. Here's an example:

> **PARENT:** Billy, I've noticed that you've been having difficulty coming in for dinner when you're playing outside. What's up?
>
> **BILLY:** You always make me come in when I'm in the middle of something fun.
>
> **PARENT:** Yes, I was thinking that's what it is. Is there anything else about my calling you in for dinner that's hard for you?
>
> **BILLY:** No. I just don't want to come in if I'm in the middle of a fun game.
>
> **PARENT:** I understand. The thing is, you're almost always in the middle of something fun when I call you in for dinner, and it's really important to me that we eat dinner together as a family.
>
> **BILLY:** I don't care if we eat dinner together as a family.
>
> **PARENT:** Um . . . OK. Well, I guess it's probably more important to me that we eat together than it is to you. But I'm thinking that if we could get the problem solved in a way that works for both of us, then we wouldn't keep fighting about it.

Your kid didn't have any ideas for solutions.

Hopefully, *you* had some ideas. Remember, it's not his job to solve the problem; it's the job of the Problem-Solving Team: you and him. So if your kid truly has no ideas, it's fine for you to offer some proposals, as long as you don't wind up imposing your will in the process. Considering solutions requires the same perseverance as drilling for concerns. (This is discussed further in Chapter 9).

Plan B never got off the ground because your kid blew up the minute you started talking or was too hyperactive to sit still for the conversation.
If your child becomes agitated the instant you try to initiate Proactive Plan B, many of the factors discussed in this chapter could be coming into play, and many of the remedies you've read about may help. Of course, there are other possible factors that could be getting in the way. For example, it's possible that your child lacks some skills crucial for participating in Plan B. That topic is covered fairly extensively in the next chapter. But there are some kids whose fuses are so short, who are so irritable and unhappy, or so hyperactive and/or inattentive, that they can't paricipate in the conversation. In those instances it's worth considering whether medication might provide some relief and make problem solving more feasible. I'm very conservative when it comes to medicating kids, so I often encourage people to see how far Plans B and C will take them before moving to medication. But there are some kids who won't be able to participate in Plan B without the aid of medication. This topic is discussed more fully in the next chapter as well.

You're too exhausted or too sick of your child to give this a try.
Here's the paradox: mustering the energy to try this approach can help you feel less exhausted and less sick of your child. Simply viewing your child's difficulties

through more accurate, more compassionate lenses can bring some energy back. Coming to a clearer understanding of your child's concerns can, too. As your child feels that his concerns are being heard, he will become more receptive to hearing your concerns, and that's going to feel good. When those draining, exhausting, demoralizing challenging episodes begin to reduce in frequency and intensity—as the problems causing those episodes are solved—you'll start putting less energy into walking on eggshells. As you become less punitive and adversarial, your kid will lash out less often. The energy and optimism will come back.

But you may need to focus yourself too. You may need to find ways to spend time away from your child to recharge and to focus on other aspects of life. Mental health clinicians, support groups, social service agencies, spouses, relatives, and friends can often be of help.

Debbie was eager to return to her Plan B conversation with Jennifer. The day after her first try, Debbie approached Jennifer during breakfast again. "Jennifer, do you remember what we were talking about yesterday morning?"

Jennifer was annoyed at being interrupted while chewing on a bite of waffle. "Yeah."

"Do you think we could finish solving the problem?"

"No."

Against her better judgment, Debbie tried again. "I was kind of hoping we could finish solving the problem."

The wooden look on Jennifer's face was a familiar one. "I'm not solving the problem."

Now things were going the way Debbie had anticipated they would the first time. She tried reflective listening. "You're not solving the problem."

"I'm not solving the problem!" yelled Jennifer, slamming her glass down on the table. "And I'm not talking about it, either!"

Debbie quickly went into de-escalation mode. "OK." She began loading dishes into the dishwasher.

After a two-minute silence, Jennifer said, "I'll talk about it later."

Debbie was tempted to ask when "later" might be but thought better of it. She decided to wait it out.

Jennifer finished her waffles, put her glass and plate in the sink, and began walking toward her bedroom. Debbie took a chance. "Let me know when you want to talk about it again."

Jennifer kept walking.

That afternoon, Debbie was talking with Kevin in the kitchen while Kevin made chili. Jennifer came into the kitchen.

"I think we should have a schedule," she announced.

Kevin, thinking she was talking about the dinner he was cooking, said, "Oh, I make my chili about every other week."

Jennifer had little tolerance for being misunderstood. "I wasn't talking about your f—ing chili!"

Kevin had little tolerance for profanity. Debbie saw where this was heading and intervened as Kevin turned to respond. "A schedule for what, honey?"

"For the TV," said Jennifer.

"For what?" said Kevin, still annoyed at Jennifer's earlier response.

"Forget about it!" yelled Jennifer.

"Whoa, hold on," said Debbie, shooting her "back-off" look at Kevin. "I want to hear your idea about the schedule for the TV."

"Not in here," said Jennifer, glaring at Kevin.

"How about in your room?" Debbie suggested. She and Jennifer settled themselves in Jennifer's bedroom. "Tell me your idea," said Debbie after they were both sitting down.

"I think there should be a schedule so me and Riley don't fight about what to watch on TV."

"Tell me more," said Debbie.

"Like, he could have a certain hour every day that he could watch *SportsCenter* and I could have an hour to watch my shows."

"I think that might be a great idea," said Debbie, who couldn't remember the last time Jennifer had proposed a solution to anything without screaming. "Shall I ask Riley if he'd be OK with that idea?"

Jennifer was silent. Debbie continued. "Because we'd want to make sure the idea works for him too."

"Well, that's my solution," said Jennifer.

"Oh, I bet he'd like the idea," Debbie reassured. "I just want to make sure."

"That's my idea, whether he likes it or not."

"Well, how about I found out if he likes it, and we can take it from there?"

Jennifer seemed finished with the conversation.

"Thanks for telling me your idea," said Debbie. "I'm glad you thought about it."

Jennifer was now distracted by her DVD player. *Looks like the conversation is over*, thought Debbie.

Debbie went back out into the kitchen. "We have a very interesting daughter," she said to Kevin.

"I don't like her swearing at us," said Kevin.

"Me either," said Debbie, sitting down at the kitchen table.

"But if I have to tolerate some swearing so she'll talk to us, I'll make that trade. Talking is more important to me right now."

"She's talking?"

"A little," smiled Debbie. "I'm starting to think there's a lot going on in that head of hers that we don't know about."

Later, it dawned on Debbie that she hadn't talked with Sandra in days. She called, excited to share the latest developments with Jennifer, but when Sandra answered the phone Debbie knew immediately that something was wrong. Sandra told her that Frankie had hit her in the mouth, left the apartment, and hadn't been heard from since. This wasn't the first time Debbie had heard that Frankie had hit his mom, and Frankie's hitting was in a different league compared with Jennifer's.

"I don't know what to do," said Sandra.

"What was he mad about?"

"I told him the new in-home therapist was coming tomorrow. He got pissed about that. Then I got pissed and told him that the in-home therapist wouldn't have to come if he'd just get his damn act together, and that he was going to get me fired from my damn job. Then he hit me. I guess he wanted me to shut up."

Debbie was at a loss. "Do you want me to come over? Do you want to meet somewhere?"

"I don't want you to see my lip."

"It won't bother me," said Debbie.

"I'll be OK." There was a long pause. "I don't want to live this way anymore," said Sandra, her voice breaking.

9

GOT QUESTIONS?

We've covered a lot of territory up to this point. And while many of your questions about solving problems collaboratively may already have been answered, it's possible many more have arisen. So this seems like a good time to pause and answer some additional questions that often come up.

QUESTION: If I'm using Plan B, how will my child be held accountable—you know, take responsibility—for his actions?

ANSWER: For too many people, the phrases "hold the child accountable" and "make him take responsibility" are really codes for "punishment." And as you read in Chapter 5, many people believe that if the punishments a child has already received for his challenging episodes haven't put an end to the episodes, it must be because the punishments didn't cause the child enough pain. So they add more pain. In my experience, behaviorally challenging kids have had more pain added to their lives than most people experience in a lifetime. If pain were going to work, it would have worked a long time ago. If a kid is putting his concerns on the table, taking yours into account, and working collaboratively toward a solution that works for both of you—and if therefore the frequency and intensity of challenging episodes are being reduced—then he's most assuredly being held accountable and taking responsibility for his actions.

QUESTION: So I can still set limits?

ANSWER: Remember, you're setting limits whether you're using Plan A or Plan B. With Plan A you're setting limits by imposing your will. You're also slamming the door on understanding and addressing your kid's concerns, increasing the likelihood of adversarial interactions, plowing ahead with uninformed solutions, not solving problems durably, and not teaching skills. With Plan B, you're setting limits by learning about what's getting in your child's way, working together on solutions that are realistic and mutually satisfactory, solving problems durably, (indirectly) teaching your child the skills he's lacking (and perhaps learning some new skills yourself), solving problems durably, and decreasing adversarial interactions. The hardest thing about setting limits using Plan B is becoming good at it.

QUESTION: Does Plan B make it clear to my child that I disapprove of his behavior?

ANSWER: Yes. He'll be crystal clear about your disapproval when you articulate your concerns in the Define the Problem step. It's also worth pointing out that a lot of the behavior you disapprove of occurs when you're using Plan A. If you stop relying on Plan A and are proactively solving problems with Plan B, much of the challenging behavior that goes along with Plan A will subside as well.

QUESTION: What about the real world? What if my kid has a "Plan A" boss someday?

ANSWER: A Plan A boss is a problem to be solved. How does your child learn to solve problems? Plan B. Which skill set is more important for life in the real world: the blind adherence to authority taught through Plan A, or identifying and articulating one's concerns, taking others' concerns into account, and working toward solutions that are realistic and mutually satisfactory taught through Plan B? If kids are completely dependent on imposition of adult will to do the right thing, then what will they do when adults aren't around to impose their will? In his book *The Global Achievement Gap: Why Even Our Best Schools Don't Teach the New Survival Skills Our Children Need—and What We Can Do About It*, my friend Tony Wagner describes the skills kids are going to need to lead productive, adaptive lives in the future. Foremost among those skills are collaboration and problem solving. Blind adherence to authority didn't make the list.

QUESTION: Aren't safety issues best addressed with Plan A?
ANSWER: It depends on the situation. Again, in emergent safety situations (e.g., your child is about to step in front of a moving car), imposition of adult will (yanking on his arm) makes perfect sense. With other emergent safety issues (e.g., your child is holding a chair over his head and threating to throw it), deescalating the situation may actually make more sense than Plan A. But, as you've read, here's the most important point: if your child is exhibiting chronic safety problems—perhaps he's *frequently* darting in front of moving cars in a parking lot—then Proactive Plan B is likely to be your best long-term option for solving that problem. Here's what that might sound like:

PARENT (INITIATING THE EMPATHY STEP): Chris, I've noticed that it's a little hard for you to stay next to me when we're in parking lots. What's up?

CHRIS: I don't know.

PARENT: Well, let's think about it a second. What's so hard about staying next to me when we're in the parking lot?

CHRIS: Um . . . I guess I'm just really excited about getting into the store.

PARENT: Yes, I've noticed that you're very excited about getting into the store. Is there any other reason you think it's hard to stay next to me?

CHRIS: Um . . . I don't like it when you hold my hand. That's for babies.

PARENT: Ah, yes, I've noticed that, too. Is there anything else you can think of that would help me understand why you're having trouble staying next to me in the parking lot?

CHRIS: Not really.

PARENT: OK. So you're having trouble staying next to me because you're really excited to get into the store and because you don't like it when I hold your hand. Yes?

CHRIS: Uh-huh.

PARENT (INITIATING THE DEFINE THE PROBLEM STEP): I understand. My concern is that it's dangerous for you to run in front of cars, and that's what happens if I don't hold your hand. And if I see that you're about to run in front of a car, I have to grab you so you don't get hurt, and then we get mad at each other. Know what I mean?

CHRIS: Yup.

PARENT (INITIATING THE INVITATION): I wonder if there's a way for us to keep you from running in front of cars in the parking lot so you don't get hurt without me holding your hand. Do you have any ideas?

CHRIS: Um . . . we could not go into parking lots.

PARENT: There's an idea. The thing is, sometimes we have to go into parking lots, like to go food shopping or to the drugstore. So I don't know if we can stay away from parking lots completely. But I bet there's some way we could be in parking lots without my having to worry about you running in front of cars and without me holding your hand. What do you think?

CHRIS: You could leave me home with Grammy.

PARENT: I could, sometimes. But Grammy can't always look after you when I'm out doing errands.

CHRIS: I could hold your belt loop.

PARENT: You could hold my belt loop. That would be better than holding my hand?

CHRIS: Yes. Holding hands is for babies.

PARENT: You'd hold my belt loop even if you were really excited about getting into the store?

CHRIS: Yes.

PARENT: What if I'm wearing something that doesn't have a belt loop?

CHRIS: Um . . . I guess I could just hold onto whatever you're wearing.

PARENT: I think that idea could work very well. Can I remind you to hold my belt loop before we get out of the car?

CHRIS: Yes.

PARENT: But sometimes you get mad when I remind you that parking lots are dangerous.

CHRIS: I only get mad if you're screaming at me to hold your hand.

PARENT: I'm screaming at you because you're . . . you know what? If you and I agree that you're going to hold my belt loop in the parking lot from now on, then it won't matter why I was screaming at you.

CHRIS: What if you forget not to scream at me?

PARENT: I'm going to try very hard not to. If I slip, can you remind me?

CHRIS: Yup.

PARENT: This plan work for you?

CHRIS: Yup.

PARENT: It works for me, too. And if our solution doesn't work we'll talk about it some more and think of another solution.

Remember, often when parents refer to "safety issues" they're referring to what their child is doing *in the midst of a challenging episode* (hitting, throwing things, etc.). Since a high percentage of challenging episodes are precipitated by Plan A, using Plan B instead of Plan A should make a big dent in the frequency of safety issues.

QUESTION: I get the importance of being proactive. But what if I should find myself in the middle of a challenging episode?

ANSWER: If you find yourself in the middle of a challenging episode, it's likely you were using Plan A. The best thing to do is defuse and de-escalate the situation so as to keep everyone safe. If you're lucky and your child is at that moment still capable of rational thought, then Emergency Plan B is an option. If not, then one option is to use Plan C at that moment and use Proactive Plan B at the next possible opportunity to solve the problem that set in motion the challenging episode in the first place. But if you have to endure a challenging episode, don't let it go to waste. Challenging episodes provide very important information about unsolved problems you may have missed or failed to prioritize. That's perhaps the only useful thing about such episodes: they let you know there's still work to be done.

QUESTION: What if I don't have time to use Plan B? It takes too long.

ANSWER: You may want to think about how long it's taking you to deal with the challenging episodes that are caused by Plan A. Most people find that challenging episodes always take longer to deal with than Plan B would have taken to prevent them. Unsolved problems always take more time than solved problems. Doing something that isn't working always takes more time than doing something that will work. If you and your child are collaborating on durable solutions, then the amount of time you're spending using Plan B will decrease over time as problems are solved.

QUESTION: I'm not that quick on my feet. What if I can't always decide what Plan to use on the spur of the moment?

ANSWER: It's only in the heat of the moment that you have to be quick on your feet. If you're solving problems proactively, you'll have to think quickly on your feet less often.

QUESTION: I started using Plan B with my daughter, and she talked! In fact, she talked so much and I gathered so much information that I started becoming overwhelmed with all the problems we need to solve! Help!

ANSWER: It's true, sometimes Plan B opens the information floodgates, and you find out there were even more problems to solve than those you identified on the *Assessment of Lagging Skills and Unsolved Problems.* While that can be overwhelming, it's good that you're now aware of all of those unsolved problems. Your goal is to add any new unsolved problems to your list, perhaps re-prioritize, and continue solving one problem at a time.

QUESTION: So I'm not a failure if I don't make it through all three steps of Plan B in one sitting?

ANSWER: Not at all! If you didn't make it past the Empathy step in your first attempt at Plan B, but you now understand your kid's concerns about that particular problem, I'd say you've been quite successful. Be sure to follow up with the next two steps before too much time passes.

QUESTION: What if my child and I agree on a solution and then he won't do what he agreed to?

ANSWER: That's usually a sign that the solution wasn't as realistic and mutually satisfactory as you may have first thought. That's not a catastrophe, just a reminder that the first solution to a problem often doesn't get the job done. Remember, effective problem solving tends to be incremental; good solutions are usually variants of the solutions that preceded them. Plan B isn't an exercise in wishful thinking. Both you and your child need to be able to follow through on the solution. If your child isn't following through, it's probably not because he won't, but because he can't. (By the way, kids aren't the only ones who don't follow through on unrealistic solutions; adults aren't very good at it, either.)

QUESTION: I did it! My child and I did Plan B together and we solved our first problem, and the solution seems to be working so far. Now what?

ANSWER: Well done. What's next? Move on to another high-priority unsolved problem, and then another. Along the way, be sure to look in the rearview mirror to take stock of the progress you're making.

QUESTION: I understand how Plan B helps me solve problems with my child. But how can I teach him the skills he's lacking?

ANSWER: Great question. The reality is that there aren't great strategies for *directly* teaching many of the lagging skills on the *Assessment of Lagging Skills and Unsolved Problems*. There is, however, a great strategy for teaching those skills *indirectly*: Plan B. When you collaboratively and proactively solve the problems that are associated with those lagging skills you are indirectly teaching those skills.

Here's an example: Let's say that Luis has difficulty making transitions (that's a lagging skill). Let's say that one example of a transition Luis is having difficulty making is turning off the TV to come to dinner. If that problem is solved through use of Plan B, several good things will happen as a result. First, the challenging behaviors that were associated with that problem will subside because the problem is now solved instead of unsolved. Second, there's now a solution in place for helping Luis make that particular transition. Is he highly skilled at making all transitions yet? No. Our goal is to improve his transition-making skills enough so that demands for transitions do not greatly heighten the likelihood of a challenging episode. When will that happen?

After we've solved some additional problems related to difficulty making transitions and Luis has more solutions in his repertoire. How will we know we're making progress? When Luis starts applying some of the solutions in his repertoire to similar transitions without a reminder or assistance. Isn't that how most kids learn new skills? Yes, it is. It's just that Luis needed a jump-start to get the developmental ball rolling—a jump-start Plan A couldn't possibly provide.

Here's another example: Because I have a busy speaking schedule, I'm a frequent flyer, so I'm often faced with flight delays or cancellations that could prevent me from reaching my intended destination in time. Now, no one has ever sat me down and provided me with direct instruction on what to do if my flight is canceled or delayed. I learned through experience, and those experiences (successful and not so successful) are the foundation of my repertoire for what to do if my flight is delayed or canceled. There are other flights on the same airline. There are other flights on a different airline. There are alternative flights to nearby destinations. There are rental cars. There are trains.

Don't most people have the skills to apply past experiences to problems they face in the present? Yes. But, but apparently not all, as evidenced by the meaningful number of fellow passengers I've seen exhibit challenging behavior when their flights are delayed or canceled.

As an added bonus, there are many skills taught just by doing Plan B, irrespective of the specific unsolved problem and associated lagging skill you're working on.

In the Empathy step, kids practice reflecting on their concerns and expressing those concerns in words without getting heated up. In the Define the Problem step, kids practice listening to another person's concerns without getting upset, listening to another person's perspective without overreacting, appreciating how their behavior is affecting others, and taking into account factors that might necessitate a change in plan. In the Invitation step, kids practice reflecting on multiple thoughts or ideas simultaneously, considering a range of solutions to a problem, considering the likely outcomes or consequences of those solutions, and shifting from an original idea or solution. Lots of skills are being taught and practiced through Plan B. And remember, it's not just the kid who's getting practice at those skills.

QUESTION: What's the role of medication in helping kids with behavioral challenges?

ANSWER: There are some kids who are so hyperactive, impulsive, distractible, irritable, or have such a short fuse and are so emotionally reactive, that it's extremely difficult for them to participate in Plan B until these issues are satisfactorily addressed. If any (or many) of these issues are making participation in Plan B difficult, then they're likely making other aspects of life difficult as well. These are issues for which medication can sometimes be helpful.

If inattention and distractibility are significantly interfering with your child's academic progress or making it difficult for him to stay focused long enough to participate meaningfully in Plan B discussions, medication may offer some promise. The mainstays of medical treatment for inattention, hyperactivity, and poor impulse control are the stimulant medications (for example, Ritalin, Focalin, Vyvanse, and Concerta), some of which have been in use for more than sixty years. Alternative, nonstimulant medications like Strattera may be appropriate for kids for whom stimulants are ineffective or who cannot tolerate their side effects.

One of the more challenging aspects of using stimulant medication is that when it's effective, many parents report that they have "two different kids": the kid who's less hyperactive and impulsive and more focused when the medication is in effect, and the kid who's the exact opposite when the medication hasn't been given or has worn off. As it relates to Plan B, this means that sometimes you have a kid who's able to sit still and focus on solving problems and sometimes you don't. It also means that when you and your kid are contemplating whether a solution is truly realistic, you must consider the possibility that solutions that are realistic when your child is medicated may not be so realistic when he's not.

Some kids are so irritable, cranky, grouchy, and grumpy that even the smallest bump in the road can feel insurmountable. A class of antidepressants called selective serotonin re-uptake inhibitors (SSRIs) such as Lexapro and Prozac may offer some relief.

Finally, if despite heavy doses of Plan B and Plan C and drastically reduced use of Plan A your child is still so short-fused or emotionally reactive that he is incapable of participating in Plan B discussions and/or severely over-reacting to infinite frustrations, a class of medications called atypical antipsychotics (like Risperdal and Abilify) may be helpful.

Many parents are opposed to medicating their child, and for good reason. Far too many kids are medicated unnecessarily, on too much medication, and taking medication for things that medication wouldn't realistically address. Medication isn't always prescribed with the necessary expertise, care, and diligence. But medication can be helpful for some of the factors contributing to challenging behavior and make it more possible for some kids to participate in Plan B. So while a conservative approach to medication is recommended, you may not want to rule it out completely. In some kids, medication is an indispensable component of treatment. It's important to remember that Plan A can heighten the potential for a challenging episode even when your kid is on an effective medication regimen.

Deciding whether to medicate one's child *should* be difficult. You'll need a lot of information, much more than is provided here, especially about side effects. Your doctor should help you weigh the anticipated benefits of medication against the potential risks so you can make an educated decision. Although it's important to have faith in the doctor's expertise, it's equally important that you feel comfortable with the treatment plan he or she proposes and the balance between benefits and risks. If you are not comfortable with or confident in the information you've been given, you need more information. If your doctor doesn't have the time or expertise to provide you with more information, you need a new doctor. Medical treatment is not something to fear, but it needs to be implemented competently and compassionately and monitored continuously.

Ultimately, what you'll need most of all is a clinically savvy, attentive, and accessible prescribing doctor, one who:

- takes the time to get to know you and your child, listens to you, and is familiar with treatment options that have nothing to do with a prescription pad;
- knows that a psychiatric diagnosis is not the most important or informative thing to know about your kid;
- understands that there are some things medication doesn't treat well at all;
- has a good working knowledge of the potential side effects of medication;
- makes sure that you—and your kid, if it's appropriate—understand each medication and its anticipated benefits and potential side effects, as well as its interactions with other medications;
- is willing to devote sufficient time to monitoring your child's progress carefully and continuously over time.

When children have a poor response to medication, it is often because one of the foregoing elements was missing from their treatment.

A discreet approach to medication is also recommended. A lot of kids don't want their classmates to know they're taking medication to address emotional or behavioral issues. If there's no way to keep the classmates from finding out, it can be helpful to have a class discussion about the other kinds of medications other children are on for other different types of conditions (asthma, allergies, diabetes) to ensure your own child does not feel singled out. On the other hand, I typically encourage parents to keep relevant school personnel well informed about their child's medication. The observations and feedback of teachers are often crucial to making appropriate adjustments in medication; the goal is to work as a collaborative team.

QUESTION: If I choose to medicate my child, how long will he be on the medication?

ANSWER: That's very hard to predict. In general, the chemical benefits of medication endure only as long as the medication is taken. However, as a child matures and develops new skills, it is sometimes possible to discontinue the medication. Ultimately, the question of whether a child should remain on medication must be revisited often.

QUESTION: What about homeopathic and natural remedies?

ANSWER: Some parents feel better about using such remedies instead of prescribed medication, and some kids benefit from them. The same standards should be applied to homeopathic and natural remedies as you would to

prescribed medication. Don't stick with it if it's not helpful, if the intervention is doing more harm than good, or
if other interventions might be more effective.

QUESTION: My child has significant language-processing and
communication delays. I'm wondering if Plan B is truly
realistic for him.

ANSWER: Since all of the examples of Plan B you've read
so far depict kids with half-decent communication skills,
it's no surprise you're wondering about this. So let's focus
for a while on how one would go about solving problems
collaboratively without the aid of the spoken word. The
good news is that these kids are already communicating;
albeit in ways that are harder to understand for those
who prefer communicating in words (that would include
most of the adults who are trying to solve problems with
them). More good news: Plan B can be adjusted for kids
with compromised communication skills so you can still
identify unsolved problems, gather some information
about the concerns related to these unsolved problems,
and help the child participate in generating and selecting
solutions.

A useful reference point, by the way, is infants. Infants may have a variety of unsolved problems: hunger,
difficulty separating or sleeping away from Mom or
Dad, difficulty eating or digesting food, difficulty es-
tablishing a regular sleep cycle, difficulty self-soothing,
difficulty dealing with the sensory world (lights,
noises, heat, cold, etc.). Although they don't have the
words to tell us about them, they *are* communicating

and we *do* collaborate with infants on solutions. We try to figure out what the infant is communicating and then apply solutions aimed at addressing the infant's concerns. And we rely on the infant's feedback to determine if the solutions have solved the problems. All without words. If we can do this with infants, we can do it with kids of any age who are unable or severely limited in their ability to communicate through use of the spoken word.

Let's think about what it would look like to engage a kid in solving problems collaboratively without the use of many (or any) words.

Identifying Unsolved Problems

The first goal remains the same: to create a list of the unsolved problems that reliably and predictably precipitate challenging episodes. While grunting, growling, and screaming are less explicit than words, they occur under specific conditions, and your observations about those conditions will help you generate a list of unsolved problems. The form the list takes depends on the extent of your child's communication skill difficulties. Here are two scenarios.

Roger is an adolescent boy who has significantly delayed expressive language skills but is able to understand much of what is being said to him. His caregivers found that his unsolved problems included being hot, being tired, feeling sick, being hungry, thinking someone was mad at him, being surprised, feeling

that people were talking too much, and having difficulty with an academic task. They wrote those unsolved problems on an index card, and whenever Roger
became agitated, they recited the possibilities to him
to find out which might be the cause. The adults soon
memorized the items, thereby eliminating the need for
the index card, and Roger eventually memorized the
list as well. In this manner, they established a beginning vocabulary for communicating about unsolved
problems. Over time, Roger became much better at
verbalizing problems. For example, instead of screaming and pounding his fists, he'd instead say, "I'm hot."
While this was certainly an improvement, most of
the problem solving still occurred in the heat of the
moment. So the adults identified the specific conditions under which each of the unsolved problems usually occurred, and then engaged Roger in thinking
about solutions *ahead of time.* At times they still had to
figure out what was troubling Roger in the heat of the
moment, but with lots of proactive solutions in place,
that need diminished greatly over time.

Of course, specific concerns such as "I'm hot" only
apply to situations in which a kid is hot. It can also be
useful to teach a more generic vocabulary of problems
that a kid can use across many situations to alert adults
that there's a problem. An especially good phrase is
"something's the matter." A kid saying *"something's the
matter"* is far preferable to having the kid *demonstrate*
that something's the matter by biting, hitting, screaming, or swearing. Teaching this phrase begins by pro-

viding the kid with direct instruction on the use of those simple words and having adults say "looks like something's the matter" whenever it looks like something's the matter. Of course, that's what's happening in the heat of the moment. It's also important for the adults to identify the specific conditions under which the kid is saying "something's the matter" so as to solve those problems proactively. The child won't need to say "something's the matter" very often once you solve the problems that are causing something to be the matter.

Adults tend to overestimate the linguistic skills we use to let people know we're frustrated or stuck or overwhelmed. The truth is, most adults lean on a few key phrases. By teaching them to kids, we're raise them to the same communication level as the rest of us.

Here's the second scenario. You met Zach in Chapter 4. He's the three-year old diagnosed with an autism spectrum disorder who uses very few words to communicate. The initial goal for Zach was the same as for Roger: to find a way to establish a basic vocabulary for unsolved problems. There are various technologies that could have helped, but Zach's speech-and-language therapist chose to use Google Images to depict, in pictures placed on a laminated card, the unsolved problems that were reliably and predictably precipitating his challenging episodes. They included being hot, being cold, being hungry, being thirsty, and something is the matter. Here's a reproduction of what that looked like:

Thirsty

Cold

Something's the matter

Hot

Hungry

When Zach needed to let his parents and teachers know there was a problem, or when he began exhibiting signs of frustration, the adults asked him to point to the picture that best communicated what was frustrating him. When Zach pointed to a picture, his caregivers verbally confirmed the problem (e.g., *"Ah, you're hungry"*). Whenever Zach encountered a problem that wasn't depicted in pictures, new pictures were added to the "problem card." After the basic menu of problems was established, a second laminated card was created that depicted potential solutions that corresponded to each unsolved problem (this is described in the next section). The long-term goal was for Zach to use words rather than point; in the meantime, he became less frustrated in his efforts to communicate with his caregivers about problems.

When a kid has significant communication challenges or other cognitive impairments it is crucial to indentify those words or concepts that are the highest priority and that need to be taught first. The words or concepts required for pinpointing unsolved problems or concerns, solving problems, and handling frustration should be a high priority because not having these words causes his most challenging moments and impedes the kid's ability to learn much else. While a vocabulary for feelings (*happy, sad, mad*) may seem important, it's more important for a child to communicate about the problems that are *causing* his sadness, anger, or frustration than to express those feelings.

Identifying and Selecting Solutions

The same strategies useful for identifying unsolved problems can be applied to identifying and choosing solutions to those problems. Zach's parents created a problem-solving binder filled with laminated cards depicting, in pictures, potential solutions for each of the problems on his "problem card."

When Zach signaled that he was hungry, he would turn to the card containing pictures of potential solutions to that problem. If it became apparent that additional solutions were needed, pictures of additional solutions were added. This binder system helped Zach communicate about both problems and potential solutions. As you'd imagine, caregivers must be consistent in their use of the problem-solving binder or it won't be as effective.

Being cold

Being hungry

Being thirsty

Being hot

Helping kids like Zach participate in the process is very important. Often it's assumed that kids with limited communication skills cannot participate in the process of solving problems, but that assumption relegates those kids to the sidelines as solutions are selected and imposed. Many such kids can, in fact, participate in Plan B, and doing so enhances relationships and fosters communication with important people in their lives. Sometimes it just takes a little extra creativity and perhaps some additional resources to get Plan B rolling. To that end, you may want to check out some of the books by Michelle Garcia-Winner and Carol Gray listed in the resources section at the end of this book.

(By the way, the problem-solving binder can be just as useful with kids whose communication skills are not compromised but who, especially in the heat of the moment, have difficulty verbalizing their concerns and thinking of potential solutions.)

A few other points before we move on. Some solutions are applicable only to certain problems. For example, a hot dog is a solution to the problem of being hungry but not for most other problems. So sometimes it's a good idea to teach a more general set of solutions. A high percentage of the solutions to problems encountered by human beings fall into one of three general categories: (1) ask for help, (2) meet halfway or give a little, and (3) do it a different way. These categories can simplify things for kids whose communication skills are compromised, who may benefit from having the three possibilities depicted in pictures, or whose communication skills are intact but who become easily overwhelmed by the universe of potential solutions.

These three categories can be used to guide and structure the consideration of possible solutions. First you'll want to introduce the categories to your child at an opportune moment. Then, when you're trying to generate solutions using Plan B, use the categories as the framework for considering solutions. As with the above examples, verbalize the words that correspond to each picture (*"Ah, do it a different way"*) to confirm the kid's idea and encourage the use of words. Then the universe of ways in which things could be "done a different way" in order to solve a problem can be explored. Let's see what that might look like:

PARENT (EMPATHY STEP, USING PROACTIVE PLAN B): I've noticed that you haven't wanted to go to gymnastics lately. What's up?

CHILD: I don't like my new coach.

ADULT: You don't like your new coach. You mean Ginny? How come?

CHILD: It's boring. All she has us do is stretch. That's boring.

ADULT: OK, let me make sure I've got this straight. You haven't wanted to go to gymnastics lately because it's boring, just a bunch of stretching.

CHILD: Right.

ADULT: Is that the only reason you haven't wanted to go to gymnastics lately?

CHILD: Uh-huh.

ADULT (DEFINE THE PROBLEM STEP): I can understand that. The thing is, you usually really like gymnastics, and you're really good at it, so I'd hate to see you give it up.

CHILD: I don't care.

ADULT: You don't care?

CHILD: Not if it's just going to be a bunch of stretching.

ADULT (INVITATION STEP): Well, I wonder if there's a way for us to do something about all that stretching without your giving up gymnastics completely. Do you have any ideas?

CHILD: Ginny's not going to change the way she does her class.

ADULT: You might be right about that. But let's think about our problem-solving options. I don't know if "asking for help" will solve this problem. And I can't think of how we would "meet halfway" or "give a little" on this one, especially if you think Ginny isn't going to change the way she does her class. I'm thinking this is one where we'd "try to do it a different way." What do you think?

CHILD: I don't know what a different way would be.

ADULT: Well, Ginny's not the only one who teaches that level. The main reason we picked Ginny's class is because the other class that's your level is the same time as your ice-skating lesson. But maybe we could change ice-skating to a different time. Then you could be in the other gymnastics class. What do you think?

This Plan B discussion would continue until a realistic and mutually satisfactory solution has been agreed on. Not only would the problem get solved, Plan B would set the stage, over time, for the child to begin using the problem-solving categories as a framework for generating solutions.

By the way, a talented speech and language therapist can take you much farther than the information provided here. It's something worth looking into.

It was Monday morning, and Debbie was in the kitchen having coffee and thinking about what had happened to Sandra the night before. Did Frankie come home? If he did, what happened then?

Jennifer came into the kitchen. "Good morning," said Debbie.

Jennifer did not respond—Debbie knew that she wouldn't — but set about the task of toasting her waffles. When the waffles were ready, Jennifer sat down to eat.

"Did Riley like my idea?" she asked suddenly.

"Sorry, what honey?" said Debbie.

"Did Riley like my idea?"

"Oh, you mean the idea about the TV schedule? I spoke with him very briefly about it last night. He seemed OK with the idea. He wasn't too sure what the schedule should be, but we didn't talk about it for very long."

"The schedule should be that I get to watch the TV when my two shows are on and he can watch before or after."

"Well, I can certainly mention that idea to him. Would you prefer that we discuss this all together, or do you want me to discuss it with you separately?"

"Separately."

"I'm picking him up at hockey practice tonight, so I can mention your idea to him then."

"Did he have any ideas?"

"Not that I'm aware of."

"Because I thought of another one, just in case."

Debbie tried to hide her surprise. "You did?"

"That's what I just said!" Jennifer said impatiently.

"Sorry, I just wanted to make sure I heard you right. What was your other idea?"

"I could record my shows, just in case he wanted to watch sometimes while my shows are on."

"That's a great idea, Jennifer. Shall I run that one by him too?"

"Yes, but I like the first idea better."

"I'll be sure to let him know that."

Jennifer's focus returned to her waffles. Debbie went back to her coffee, glancing occasionally at her daughter, in slight disbelief. Jennifer had revisited the discussion! She'd come up with more than one solution! She wanted to know if her brother was OK with her idea! Debbie couldn't resist the temptation . . . she got up and gave Jennifer a quick hug.

This did not go over well. "Why'd you do that?!" shouted Jennifer, immediately pushing Debbie away and stalking off to her room with her waffles. But Debbie thought she noticed the slightest hint of a smile on Jennifer's face as she departed.

"My partner," Debbie whispered when Jennifer was gone. "My problem-solving partner."

That night, Debbie grew increasingly concerned that Sandra wasn't answering her phone or responding to texts. Sandra finally called at around 9:30 p.m., sounding harried. Debbie understood why when Sandra recounted her day. Worried that Frankie hadn't come home all night, Sandra had paged Matt, the new home-based mental health counselor. He'd come over and encouraged Sandra to call the police to help find Frankie. Just as she was about to, Frankie walked in. Sandra was sure Frankie would go ballistic when he saw Matt in the apartment, but he didn't. "He saw my lip and got really remorseful," said Sandra.

Sandra related that Matt was able to get Frankie to start talking. Frankie told him that he was sorry he'd hit her, that he hates his new program at school, that the program staff are mean and the kids are way more screwed up than he is, and that his new medicine is making him feeling really jittery.

"And Frankie told Matt that he's really sick of being trouble all the time and that he's been smoking a lot of weed at his friend Tyler's—that's where he was last night—because that's the only thing that makes him feel better. And that he's really scared."

"Scared?" asked Debbie.

"Yeah, scared. He said he feels out of control, and like there's no one who can help him."

"Unbelievable."

"That's what I'm sitting there thinking. I mean, the kid hasn't talked this much for like five years."

Frankie had nodded when Matt asked if he was thinking about hurting himself. Then Frankie said he didn't want to talk in front of Sandra anymore, so Sandra sat in her bedroom while Matt and Frankie talked. About ten minutes later, Matt came to her and said he thought Frankie needed to be in the hospital.

"Oh, no," said Debbie.

"That's what I'm thinking!" said Sandra. "But Matt says Frankie's on board with the idea because there's an inpatient unit over in Amberville where they don't restrain kids or throw them in seclusion rooms. So me and Matt and Frankie got in Matt's car and we drive over to the inpatient unit and that's where Frankie is now."

"So he's still there?"

"Yeah, he might be there for a week."

"I'm so sorry you had to go through all that," Debbie said. "Are you OK?"

"I'm glad he's safe," said Sandra. "So I'm kind of relieved." She paused. "But I'm really sad that he was going through all of that and couldn't tell me. I wish . . ." Sandra couldn't continue.

"Maybe this will be what gets things on track," said Debbie.

Sandra tried to collect herself. "I'm not getting my hopes up. We've been through this before. We'll probably be right back at square one when he gets out."

"Maybe this time it'll be different."

When the conversation ended, Debbie sat quietly by the phone. She felt like crying, but wasn't able to figure out why. She couldn't decide whether to feel hopeful or helpless. How did Sandra find the strength to deal with everything she had on her plate? Why hadn't she just thrown in the towel a long time ago? Then she answered her own question. "Because it's her kid," she whispered quietly. "When it's your kid, you keep going." It was good that Frankie was finally talking to someone. She hoped the people at the hospital knew what they were doing. She wondered about the things that went on in Jennifer's head, things she knew nothing about. Still, after so many years of worrying and arguing and screaming, she felt the slightest glimmer of hope that she was finally getting to know that prickly, closed-book daughter of hers. She sighed. "Why does this have to be so hard?"

Debbie stood up and began walking toward her computer—maybe that inpatient unit in Amberville had a website—but then she did an abrupt about-face, bumping into Kevin, who was headed toward the refrigerator.

"Where you going?" Kevin asked.

"I'm not sure," said Debbie, "but I was just thinking of going to see if your daughter wants me to tuck her into bed."

"She hasn't wanted you to tuck her in for years."

"I know," said Debbie. "I think that's because I've been so caught up in who she *isn't* that I've been blowing right past who she *is*. I've let that get in the way of the most important parts of being her mother. And I don't want it to be that way anymore."

10

FAMILY MATTERS

Every family has its challenges. Siblings don't always get along, parents don't always see eye to eye, everyone's too busy, kids are stressed about school or grades or friends, adults are stressed about work or money or trying to carve out time for themselves, and just about everyone's stressed about homework.

Add a behaviorally challenging child to the mix, and many families and marriages will be pushed to the brink. Small annoyances turn into big problems, minor disagreements and stressors become major catastrophes, and communication problems that might never have been noticed become glaring roadblocks. Now add grandparents who remember the way they would've done things in the good old days and soccer or hockey coaches or teachers who are delighted to tell you how they'd handle your kid, and life has suddenly become much more challenging and demanding than you ever expected.

While all of that may feel overwhelming at times, we're going to deal with all those folks the same way we've been trying to handle the problems that need to be solved with your behaviorally challenging child: one at a time.

SIBLINGS

A behaviorally challenging child can make run-of-the-mill sibling rivalry look like a walk in the park. While it's not uncommon for "ordinary" siblings to direct their greatest hostility toward one another, these acts can be more intense, frequent, and traumatizing when one of the siblings is behaviorally challenging. And while it's not unusual for "ordinary" siblings to complain about preferential treatment and disparities in parental attention and expectations, these issues can be magnified in families with a behaviorally challenging child because he may require such a disproportionate share of the parents' resources.

So we have a few important things to consider:

- We want to protect siblings from the verbal and physical aggression of their behaviorally challenging brother or sister, but we also want to recognize that though the challenging child's response to difficult sibling interactions is way over the top, those interactions involve two people. In other words, it takes two to tango.

- We want to make sure that siblings get the attention and time and nurturing they need from you, but also understand that the challenging kid may require a disproportionate share of that attention and time and nurturing, at least until you get some problems solved and some skills taught.
- We want to make sure that siblings understand why their brother or sister is reacting in such a powerful way to difficulties and disagreements and having so much trouble being flexible, tolerating frustration, and solving problems, but we don't want them to feel like they're walking on eggshells or that their needs and concerns always take a backseat to the seemingly more pressing needs of the challenging child.
- We want to make sure that the siblings know that we appreciate how hard it is to have a behaviorally challenging brother or sister, but we need to empathize in a way that isn't disrespectful or dismissive of the challenging child's genuine and significant needs.

You have your work cut out for you. Fortunately, you have Plan B, and it's just as applicable to solving problems between siblings as it is to solving other problems.

If they are old enough to understand, it is often useful to explain to brothers and sisters why their behaviorally challenging sibling acts the way he does, why his behavior is so difficult to change, how to interact with him in a way that reduces hostility and minimizes the likelihood of aggression or explosions, and the steps you're taking to improve things. Brothers and sisters tend to be more receptive if there's an improvement in the general tone of family interactions and if their behaviorally challenging sibling blows up less often and is an active participant in making things better.

However, this understanding doesn't always keep siblings from complaining about an apparent double standard between themselves and their behaviorally challenging brother or sister. Armed with the knowledge that parental attention is never distributed with 100 percent parity and that parental priorities are never exactly the same for each child in any family, you should resist responding by trying even harder to treat your behaviorally challenging child the same as you do your other children. In all families—yours and everyone else's—*fair does not mean equal.* Even parents in "ordinary" families often find themselves helping one child more with homework, having higher academic expectations for one child, or being more nurturing toward another. In your family, you're doing things a little differently for the child who needs extra help in the areas of flexibility, frustration tolerance, and problem solving, but you're also doing things differently for the other children, who have challenges of their own. When siblings complain about disparities in parental expectations, it's an excellent opportunity to empathize and educate.

RILEY: How come you don't get mad at Jennifer when she swears at you? It's not fair.

DEBBIE: I know that it's hard for you to listen to her swearing. I don't like it very much, either. But in our family we try to help one another and make sure everyone gets what they need. I'm trying to help Jennifer solve some frustrating problems and to help her think of different words she can use instead of swearing. That's what she needs help with.

RILEY: But swearing is wrong. You should get mad at her when she swears.

DEBBIE: Well, I don't get mad at you when I'm helping you with your math, right? That's because I don't think getting mad at you would help very much. Remember how I used to get mad at Jennifer whenever she swore? It didn't work very well, did it? It just made things worse. So I'm doing something now that I think will eventually work better. I think it's starting to work pretty well.

RILEY: What are you going to do if I start swearing?

DEBBIE: I'd help you think of different words, too. Then again, you don't seem to have a problem with swearing, which is really good. So it doesn't look like that's what you need my help with.

What should you do to resolve disputes and disagreements between siblings? Apply Plan B. The ingredients are the same, but your role is now that of Plan B facilitator. You'll still want to take it one problem at a time. Because problems between siblings tend to be highly predictable, Proactive Plan B is still far preferable to Emergency Plan B. You'll want to ensure that the concerns of both siblings are identified and clarified. (Often this is best accomplished by doing the Empathy step in separate discussions with both siblings prior to bringing them together to discuss potential solutions.) And you'll still want to ensure that the agreed-on solutions truly address the concerns of both parties and are realistic and mutually satisfactory.

Over time, siblings of behaviorally challenging kids feel better when problems are resolved through Plan B because they see that their concerns are being heard, understood, and taken into account. They come to see their behaviorally challenging sibling as more approachable and less terrifying. They appreciate being involved in the process of working toward solutions and recognize that you're able to handle the process in an evenhanded manner.

Here's what Plan B between two siblings looks like, with a parent as facilitator:

Preliminary Plan B with Sibling #1 (Sibling #2 is not present):

PARENT: I've noticed that you and your brother are fighting a lot when you're in the playroom together. What's up?

ANDREW: Caleb always plays with my toys.

PARENT: Ah, so you don't want him playing with your toys. But I thought we were keeping your toys in your bedroom and his toys in his bedroom, so I thought the toys in the playroom were for sharing.

ANDREW: Right.

PARENT: So I don't think I understand what you mean when you say "your" toys.

ANDREW: The ones I'm playing with.

PARENT: Ah, so Caleb tries to play with the toys that you're still playing with.

ANDREW: Uh-huh.

PARENT: Does he know you're still playing with them?

ANDREW: I don't know. He doesn't ask me.

PARENT: How would he be able to tell you're still playing with them?

ANDREW: I don't know.

PARENT: Can you give me an example of a toy that you might be playing with and then he starts playing with it?

ANDREW: The cars.

PARENT: Ah, the cars. So you'll be playing with the cars and then he'll butt in and want to play with them too?

ANDREW: Well, I'm not exactly playing with them. But I'm not done with them yet.

PARENT: Oh, I see. So you're not still using them but you're also not done with them. Yes?

ANDREW: Yes.

PARENT: How would Caleb know you're not done with them if you're not still using them?

ANDREW: I don't know.

PARENT: And how much time should pass when you're not playing with them before you're through with them?

ANDREW: I don't know.

PARENT: OK, I think I understand. I'm going to talk to Caleb about this, too, because fighting over the toys is making you guys hurt each other and that's not okay in our house.

ANDREW: OK.

Preliminary Plan B with Sibling #2 (Sibling #1 is not present):

PARENT: Caleb, can we talk a little about what's going on between you and Andrew when you guys are playing together?

CALEB: OK.

PARENT: Why do you think you guys are fighting so much?

CALEB: He won't let me play with the toys I want to play with.

PARENT: How come he won't let you play with the toys you want to play with?

CALEB: He says he's still playing with them. But he's not still playing with them. And then there's nothing for me to play with!

PARENT: So he gets mad if you play with toys that it seems like he's not playing with anymore.

CALEB: Yes!

PARENT: So you're not trying to play with the toys he's playing with at that moment?

CALEB: No, I'm trying to play with something else. But he says he's still playing with everything I try to play with.

PARENT: So there's nothing left to play with.

CALEB: Uh-huh. Then he hits me when I try to play with something.

PARENT: We need to solve this problem, don't we?

CALEB: Yes, because I never get to play with anything if Andrew's around.

PARENT: I think we need to have a meeting with Andrew so we can talk about it.

Plan B with Siblings #1 and #2:

PARENT: I've talked with both of you about the problem you've been having playing with toys together, and I thought it would be good to come up with a solution together. Andrew, you told me that sometimes you're still playing with toys even though you're not exactly using them, yes?

ANDREW: Yes.

PARENT: And Caleb, you told me that there's so many toys that Andrew is still playing with that there's nothing left for you to play with, yes?

CALEB: Uh-huh.

PARENT: And I'm concerned because when you guys are having trouble sharing toys, you end up hitting each other and someone gets hurt, and that's not the way I want people to treat each other in our family.

CALEB: He hits me first!

ANDREW: I wouldn't hit you if you didn't touch stuff I'm still playing with!

PARENT: Um, I don't know if we'll ever figure out who hits who first. I do think we can solve this problem, though, so no one's hitting anyone. I wonder if there's a way for Caleb to know what toys you're still playing with, Andrew, but still have some toys to play with himself. Do you guys have any ideas?

ANDREW: He could stay out of the room I'm playing in.

PARENT: Well, that's one idea. But if you're in the play-room and Caleb isn't allowed in there while you're in there, I don't know if that would be fair to Caleb.

ANDREW: But he has toys in his room. He could play with them. And then he wouldn't touch mine.

CALEB: I don't want to play with the toys in my room all the time. I want to play with the toys in the playroom sometimes.

PARENT: So, Andrew, let's hold on to that solution just in case we can't come up with anything else. Any other ideas for how we could know what toys Andrew is still playing with but still have some toys for Caleb to play with?

ANDREW: I could tell him what toys I'm still playing with.

CALEB: You already do that . . . and it's everything.

ANDREW: Well, I'm not through playing with everything yet.

PARENT: Andrew, how long does it take for you to be done playing with something?

ANDREW: I don't know.

PARENT: Like, we're sitting here talking right now. And you haven't been in the playroom since this morning. Is there anything you're still playing with in the playroom?

ANDREW: Um . . . the cars.

CALEB: No way! He hasn't been in there since this morning.

ANDREW: Yeah, but I have them set up a certain way and I don't want you to wreck 'em.

PARENT: So I wonder what we could do about this. Caleb feels it's not fair if you're never through playing with the cars. And Caleb, Andrew would prefer that you not play with the cars if he has them set up a certain way and doesn't want you to wreck the setup. This is a hard one!

CALEB: At school, you're done playing with a toy when playtime ends.

PARENT: Hmm. So when playtime ends, it's a fresh start on who's playing with the toys?

CALEB: Uh-huh. That's how it is at school. But not here.

PARENT: Well, maybe it could work here. Andrew, what do you think of the idea of having a time limit on how long you're still playing with toys that you haven't used in a while, like they do at school?

ANDREW: How long?

PARENT: I don't know. That's for you guys to decide. I'm wondering what you think of the idea.

ANDREW: Maybe it could work.

CALEB: I think he should be done playing with a toy as soon as he's not using it anymore.

PARENT: What do you think, Andrew?

ANDREW: That's too quick.

PARENT: Any ideas for what wouldn't be too quick?

ANDREW: Ten minutes. If I haven't used a toy for ten minutes, I'm through playing with it.

PARENT: Caleb, what do you think?

CALEB: That would give me a lot more toys to play with.

PARENT: Andrew, this could be very hard for you. Caleb would be able to play with the cars you have set up right now because it's been a lot longer than ten minutes since you used them. Can you do that?

ANDREW: Maybe Caleb would listen to me if I asked him not to play with the cars because I have them set up . . . but he could play with everything else.

PARENT: Caleb, could you do that?

CALEB: Yes, if he told me. But he doesn't tell me. He just tells me I can't play with anything.

PARENT: So let's think about what we're deciding here. Andrew, if you haven't used something for ten minutes, then you're through playing with it. And Caleb, if Andrew tells you that he's set something up in an extra-special way, then you'll try not to play with it. Yes?

ANDREW: Yes.

CALEB: Uh-huh.

PARENT: Well, we'll have to see how this solution works. If it doesn't, don't start hitting each other; just let me know so we can keep working on it.

These discussions can require a meaningful amount of adult guidance and management. Early on, siblings may need help listening to each other, waiting for each other to finish talking, taking turns, not overreacting to observations and ideas with which they don't agree, and so forth. If they can't handle being in the same room for the Invitation step, caregivers sometimes need to do some "shuttle diplomacy"—going back and forth between the two siblings without any face-to-face discussions between them—until some problems have been solved and the siblings have some practice and faith in the process.

One other thing to watch out for: in some instances, the behavior of seemingly angelic siblings can begin to deteriorate just as the behavior of their behaviorally challenging brother or sister begins to improve. This is often a sign that the emotional needs of the siblings, which had been below the radar while the family dealt with the pressing issues of the behaviorally challenging child, require closer attention. In some cases, therapy may be necessary for brothers and sisters who have been traumatized by their behaviorally challenging sibling or who begin to manifest other problems that can be traced back to the old family atmosphere.

If you feel that your family needs more help working on these issues than this small section provides, a skilled family therapist can be of great assistance. You may also wish to read an excellent book, *Siblings Without Rivalry*, by Adele Faber and Elaine Mazlish.

COMMUNICATION PATTERNS

A family therapist can also help when it comes to making some fundamental changes in how you communicate with your child. Dealing effectively with a behaviorally challenging kid is easier (not easy, *easier*) when there are healthy patterns of communication between him and his parents. When these patterns are unhealthy, dealing effectively with such a child is much harder. While some of these unhealthy patterns are more typical of older behaviorally challenging kids, the seeds for them may be sown early on. Although it's not an exhaustive list, here's a sampling of some of the more common patterns.

Parents and children sometimes get into a vicious cycle of drawing erroneous conclusions about each other's motives or thoughts. This pattern may be referred to as "speculation," "psychologizing," or "mind reading," and it can sound something like this:

PARENT: The reason Oscar doesn't listen to us is that he thinks he's so much smarter than we are.

It's common for people to make inaccurate inferences about one another, and responding effectively to these inaccuracies—in other words, setting people straight about yourself in a manner they can hear and understand—is a real talent and requires some big-time emotion regulation and communication skills. While some kids can respond to speculation by making appropriate, corrective statements to set the record straight ("Dad, I don't think that's true at all"), a behaviorally challenging kid may hear himself characterized inaccurately and become extremely frustrated. This is an undesirable circumstance in and of itself, especially because whether Oscar thinks he's smarter than his parents isn't really the point. In fact, this topic is a detour that distracts everyone from working collaboratively toward solutions to the unsolved problems that lead to Oscar's challenging episodes. Of course, speculation can be a two-way street. From a child's mouth, it might sound something like this:

OSCAR: The only reason you guys get so mad at me so much is because you like pushing me around.

When adults respond in kind, the detouring simply continues:

MOTHER: Yes, that's exactly right: our main goal in life is to push you around. I can't believe you'd say that, after all we've been through with you.

OSCAR: Well, what is your main goal then?

FATHER: Our main goal is to help you be normal.

OSCAR: So now I'm not normal. Thank you very much, loser.

FATHER: Don't you get disrespectful with me, pal.

Speculation is a no-win proposition. Solving problems collaboratively is a win-win proposition. So let's stick to the script and, rather than *speculating* about what another family member is thinking or feeling, we'll *drill* for that information instead. That takes a lot of the guesswork out of the mix. In the Empathy step, you're trying to get your kid's concerns on the table. In the Define the Problem step, it's your turn. No psychologizing. No mind reading. No value judgments. Just concerns.

"Overgeneralization"—drawing global conclusions about isolated events—is another maladaptive communication pattern. Here's what it would sound like when a parent does it:

MOTHER: Ernesto, can you please explain to me why you never do your homework?

ERNESTO: What are you talking about? I do my homework every night!

MOTHER: Your teachers told me you have a few missing assignments this semester.

ERNESTO: So does everybody! What's the big deal? I miss

a few assignments, and you're ready to call in the damn cavalry!

MOTHER: Why do you always give me such a hard time? I just want what's best for you.

ERNESTO: Stay out of my damn business! That's what's best for me.

There may be ways Ernesto's mother could help him with his homework or at least get some of the reassurance she wants that he'll complete his homework. But not by starting the discussion with an overgeneralization. While other children are sometimes able to bypass their parents' overgeneralizations and get to the real issues, many behaviorally challenging kids often react strongly to such statements and lack the skills to respond appropriately with corrective information. You're best off phrasing things as an unsolved problem ("Ernesto, your teachers tell me you're missing a few homework assignments . . . what's up?") and leaving the overgeneralizations on the shelf.

Another common tendency—"perfectionism"—sometimes prevents parents from acknowledging the progress their child has made while they cling to an old, unrealistic vision of the child's capabilities. Perfectionism is usually driven less by the child's lack of progress and more by the parents' own anxiety. Wherever it's coming from, perfectionism is counterproductive when applied to a child tired of receiving feedback on practically everything he does or who feels frustrated by his parents' unrealistic expectations:

FATHER: Eric, your mother and I are pretty pleased about how much better you're doing in school, but you're still not working as hard as you ought to be.

ERIC: Huh?

MOTHER: We think you should be working harder.

ERIC: I get my work done, don't I?

FATHER: Yes, apparently you do, but we want you to do extra math problems over the weekend so you can get even better at it.

ERIC: Extra math problems? I already have too much homework over the weekend.

FATHER: Well, that may be true, but we really think the extra math practice will be very helpful to you.

ERIC: I'm not doing extra math problems over the weekend. I need a break over the weekend.

MOTHER: We're just trying to look out for you. Now, your father and I have already talked this over, so there's no discussion on it.

ERIC: No freaking way.

Eric may or may not actually be interested in thinking about how to improve in math. Either way, his parents would be far better off approaching him about it through Plan B.

Here are some other maladaptive communication patterns you want to avoid:

SARCASM: Sarcastic remarks are often totally lost on behaviorally challenging kids (especially the black-and-white thinkers) because they don't have the skills to figure out that the parent means the exact opposite of what he or she is actually saying.

PUT-DOWNS: These are a poor way to engage a kid in solving problems collaboratively. (*"What's the matter with you? Why can't you be more like your sister?"*)

CATASTROPHIZING: Greatly exaggerating the effect of current behavior on a child's future well-being does no one any good. (*"We've resigned ourselves to the fact that James will probably end up in jail someday."*)

INTERRUPTING: Don't forget that your child is probably having trouble sorting out his thoughts in the first place. Your interruptions don't help.

LECTURING: *"How many times do I have to tell you . . . ?"* You've probably told him more than enough times, so it's better to switch gears and try to figure out and resolve whatever is getting in the way of his doing what you've been telling him to do.

DWELLING ON THE PAST: *"Listen, kid, your duck's been upside down in the water for a long time. You think I'm gonna get all excited just because you've put together a few good weeks?"* Ouch.

TALKING THROUGH A THIRD PERSON: *"You're not going out with your friends this weekend, and your father is going to tell you why. Isn't that right, dear?"* Whether you're hiding behind someone else or not, Plan A is not the ideal way to get your concerns addressed.

All these communication patterns are very common, and very counterproductive.

Your goal is to change these patterns, identify unsolved problems proactively, effectively drill for information, understand your child's concerns, resist the temptation to dismiss them, articulate your own concerns, and patiently consider and evaluate potential solutions without veering into Plan A. This is very hard to do but very worth the effort. There's no shame in having difficulty doing it on your own; seek out a reputable family therapist if you need help. Make sure they know a thing or two about solving problems collaboratively.

You probably remember Mitchell, one of the kids you met in Chapter 4. Let's return to a therapy session in which his family's communication patterns were being discussed.

Mitchell and his parents arrived for their second meeting with their new therapist, who was advised that it had been a difficult week.

"We can't talk to him anymore—about anything—without him going crazy," said Mitchell's mother, Kathryn.

"THAT'S NOT SO, MOTHER!" Mitchell yelled. "I'm not going to sit here and listen to you exaggerate."

"Why don't you stand then?" cracked Paul, Mitchell's father.

Mitchell paused, reflecting on his father's words. "If you were joking, then you're even less funny than I thought you were. If you weren't, then you're dumber than I thought you were."

"I'm not the one who flunked out of prep school," Paul jabbed back.

"AND I'M NOT THE ONE WHO MADE ME GO TO THAT SCHOOL!" Mitchell hissed.

"Look, I'm really not interested in getting into a pissing contest with you, Mitchell," said Paul.

"What do you call what you just did?" Kathryn chimed in. "Anyway, I don't think Mitchell is ready to face flunking out of prep school yet."

"DON'T SPEAK FOR ME, MOTHER!" Mitchell boomed. "YOU DON'T KNOW WHAT I'M READY TO FACE!"

"Pardon me for interrupting," the therapist said, "but is this the way conversations usually go in this family?"

"Why? Do you think we're all lunatics?" asked Mitchell.

"Speak for yourself," said Paul.

"Screw you," said Mitchell.

"Well, we're off to a wonderful start, aren't we?" said Kathryn.

"WE ARE NOT OFF TO A WONDERFUL START, MOTHER!" Mitchell yelled.

"I was being sarcastic," said Kathryn. "I thought a little humor might lighten things up a bit."

"I'm not amused," Mitchell grumbled.

"Fortunately, we're not here to amuse you," said Paul.

"Sorry to interrupt you folks again," the therapist said, "but I'm still wondering if this is a pretty typical conversation."

"Oh, Mitchell would have gotten insulted and stormed out of the room if we were at home," said Kathryn. "In fact, I'm surprised he's still sitting here now."

"YOU HAVE NO IDEA HOW I FEEL!" yelled Mitchell.

"We've been listening to you telling us how you feel since you could talk," said Paul. "We know more about how you feel than you know."

"ENOUGH!" yelled Mitchell.

"My sentiments exactly," the therapist said. "I think I'll answer my own question. Forgive me for being so direct, but you guys have some not-so-wonderful ways of communicating with one another."

"How do you mean?" asked the mother.

"You're a very sarcastic group," the therapist said. "Which would be fine, I guess, except that when you're sarcastic, I think it makes it very hard for Mitchell to figure out what you mean."

"But he's so smart and we're so dumb," said Paul.

Mitchell paused, reflecting on his father's words. "Are you trying to be funny again?" he asked.

"You're so smart, figure it out," Paul jabbed back.

The therapist interrupted, "I'm sure you guys could do this all day, but I don't think it would get us anywhere."

Mitchell chuckled. "He still thinks we're going to accomplish something by coming here."

"I should add that sarcasm isn't the only bad habit," the therapist continued. "The one-upmanship in this family is intense."

"Birds of a feather," Kathryn chirped.

"What does that mean?" Mitchell demanded.

"It means that the apple didn't fall far from the tree," said Kathryn.

"Be careful about whose tree you're talking about," said Paul. "I don't want any credit for this."

"Oh, I'm afraid you're right in the thick of things," the therapist reassured the father. "I wonder if we could establish a few rules for communicating. I must warn you, I'm not sure you'll have much to say to one another once I tell them to you."

"We wouldn't have to talk to each other?" Mitchell said. "Excellent."

"What kind of rules?" asked Kathryn.

"Well, it would be a lot more productive if we got rid of a lot of the sarcasm," the therapist said. "It really muddies up the communication waters. And the one-upmanship has got to go."

Paul broke the ensuing silence. "I don't think he can do it," he said, looking at Mitchell.

Before Mitchell could erupt, the therapist interjected, "That's one-upmanship."

Mitchell's frown turned upside down. "Thank you," he said.

"Ooh, this guy is tough," said Paul, turning to his wife. "I don't like coming here anymore." He smiled.

"My husband doesn't like being corrected, especially in court," said Kathryn.

"Fortunately, I'm not a judge . . . or jury," said the therapist.

"Yeah, well, he's in attorney mode pretty much full-time," said the mother. "He thinks everything's all about posturing and leverage and manipulating rules of evidence to shade the truth. Finding out what's really true isn't exactly the goal."

"Truth is over-rated," said Paul.

"Oh, there's another bad habit I should mention," the therapist said.

"Oh, no, what did I say?" Kathryn said, covering her mouth.

"You guys talk for one another a lot," the therapist said, "like you can read one another's minds."

"Well, we know each other very well," said Kathryn.

"That may be," the therapist said, "but from what I've observed, your speculations about one another are often off-target, and they don't go over well."

"What'd you call it?" asked Kathryn.

"Speculation," the therapist said. "Thinking you know what's going on in someone's head. It just gets you guys more agitated with one another."

"No more speculation?" said Kathryn.

"Not if you guys want to actually start talking to one another," the therapist confirmed.

"What should we do if someone does one of those three things?" Mitchell asked.

"Just point it out to them without being judgmental," the therapist said. "If someone is sarcastic, just say, 'That's sarcasm.' If someone is one-upping, say, 'That's one-upmanship.' And if someone is speculating, say . . .'"

"'That's speculation,'" Mitchell interrupted.

"My, we catch on fast," said Paul.

"That's sarcasm," said Mitchell.

GRANDPARENTS

At times it's necessary to bring grandparents into the loop on understanding, and perhaps helping, a behaviorally challenging child. In many families, grandparents or other relatives function as co-parents, taking care of the children while the parents are at work. Even if grandparents don't spend much time with the child, they benefit from learning about the lagging skills and unsolved problems that set the stage for their grandchild's challenging behavior. If they never miss an opportunity to tell parents what they would do if *they* were in charge, it will benefit you *and* them if they understand that the way things were done in the good old days doesn't teach lagging skills or solve problems durably. I've seen situations in which grandparents played an absolutely indispensible role in helping a behaviorally challenging kid because it was a grandparent who had the best relationship with the kid and was therefore in the best position to begin the process of solving problems collaboratively.

YOU

A behaviorally challenging child can put tremendous strain on a marriage. In many two-parent families, one parent is primarily disposed toward imposition of adult will, convinced that more authority and firmer limits are what the child needs most, while the other chooses to let things go, convinced that more authority is only making things worse. Since neither approach is working, they have little to show for their predispositions, and it's not unusual for the two adults to blame each other for the failure to make much headway in reducing challenging episodes:

> **PARENT #1:** If you'd just let me deal with him and stop letting him off the hook, things would be different around here.

> **PARENT #2:** I'm not going to stand by and watch you scream at him and punish him all the time. Somebody needs to give the kid a break!

Just as behaviorally challenging kids can cause tension between caregivers, significant tension between caregivers can make life with a behaviorally challenging child much more difficult. Some partners aren't very good at collaboratively solving problems with each other, so working on unsolved problems with a child can require new skills. Partners who are drained by their own difficulties often have little energy left for a labor-intensive child with behavioral challenges. Sometimes one partner feels exhausted and resentful that he or she has to be the primary caregiver because the other spends a lot of time at work.

It's hard to work on helping your child if you're feeling the need to put your own house in order first. Perhaps you've realized that you're lacking some of the same skills as your behaviorally challenging child (this may have become apparent as you were perusing the lagging skills on the ALSUP). Plan B can help you learn new skills right along with your child. Perhaps your challenging child's behaviors evoke strong emotions in you because those behaviors are reminiscent of abusive experiences you've endured yourself. Plan B can help you reduce those behaviors and feel OK about pursuing expectations and solving problems with your child in a way that bears no resemblance to those past experiences.

Perhaps you're so drained by work and schedules and the needs of your other children that you simply have little energy and patience left for the rigors of helping your behaviorally challenging child. Plan B can help you get your energy back. Solved problems aren't energy-drainers, only unsolved problems are. Maybe you're bitter about having been dealt a difficult hand. Plan B can help you be more responsive to the hand you've been dealt and play those cards well. Finally, maybe you feel that you need to get a better handle on your own temper so you can help your child do the same. You should find the proactive and collaborative aspects of Plan B to be helpful.

These things don't change on their own. Make sure you take care of yourself. Working hard at finding or creating a support system of your own is essential. Seek professional help or other forms of support if you need it.

Q & A

QUESTION: My spouse won't use Plan B. He won't even read this book. Any advice?

ANSWER: Lots of adults use Plan A out of sheer habit. They actually may not have strong beliefs guiding their use of Plan A, it's just what they were raised with, or it's what a lot of books and talk-show hosts and nanny programs tell them to do, and they've never given the matter much thought. The goal is to help them give the matter some thought, beginning with new information about a child's lagging skills and the association between those lagging skills and the child's unsolved problems. Hopefully, your spouse will come to the recognition that your child and family will be helped far more if the adults view themselves as problem-solvers rather than as the swift and unrelenting purveyors of adult-imposed consequences.

For some adults, reading a book isn't the preferred way to access new information. Maybe your spouse will listen to a CD in the car or visit a web site. Check out www.livesinthebalance.org for lots of free resources.

There are also many adults who rely on Plan A because they fear that their concerns won't be heard or addressed if they attend to their child's concerns. When did they come to fear that their concerns wouldn't be heard of addressed? Probably during childhood, when their Plan A parents neither heard nor addressed their concerns. These adults need to be assured that their concerns will be heard and addressed using Plan B as well.

QUESTION: What if my spouse says Plan A worked for him when he was a kid?
ANSWER: There are adults who feel that being raised on Plan A contributed positively to their development. And maybe your spouse did have skills to handle imposition of adult will. But your behaviorally challenging child is not handling imposition of adult will very well at all. So what apparently "worked" for your spouse clearly isn't "working" for your behaviorally challenging child.

QUESTION: I've been taught that it's important for parents to have a united front so the kids can't do any "splitting" and play one parent against the other. But what if there are significant areas of disagreement on child rearing between me and my co-parent?
ANSWER: It sounds like you and your co-parent are already "split," so you may be giving your kid more credit for that outcome than he deserves. Hopefully this book will help you and your co-parent start agreeing on a few key points: (1) the unsolved problems that you're going to be working on and the ones you're setting aside for now;

and (2) how you want to go about solving the problems you've decided you're working on.

QUESTION: My other children are not especially challenging and respond well to Plan A. Can I have two different types of discipline going on in my household at the same time?

ANSWER: Kids who respond to Plan A tend to respond to Plan B as well, so if you're determined to be consistent, use Plan B with your not-so-challenging kids, too. But there isn't a household in the world where all the children are treated exactly the same. In all households, one child is getting something another isn't getting. Your not-so-challenging kids want your challenging child to stop having challenging episodes far more than they want everyone to be treated exactly the same.

After work, Sandra drove to the inpatient unit for her first meeting with the unit social worker, Ms. Brennan. Frankie had been on the unit for two days.

"That lip looks pretty bad," said Ms. Brennan.

"It looks worse than it feels," said Sandra, lying.

"From what I can gather, you and Frankie have been down this road before," said Ms. Brennan.

"You mean him hitting me or him being on an inpatient unit?"

"Well, both, I guess."

"I've been hit many times," said Sandra. "Not usually this bad. He's always sorry afterward. We don't really talk to each other much anymore, so I haven't been getting hit much lately. And, yeah, he's been hospitalized a few times before this. It's never really accomplished anything. To tell you the truth, I think the other inpatient units did more harm than good."

"I'm sorry to hear that," said Ms. Brennan. "I thought I'd talk with you for a few minutes before we bring Frankie in on the discussion. Tell me about your life with Frankie."

"It's been hard," said Sandra.

"Tell me about that."

"I was sixteen years old when I had him. We lived in a homeless shelter when he was younger. But we did OK back then. Now, like I said, we don't really communicate much anymore. And when we do, it's ugly."

"I understand," said Ms. Brennan.

No you don't, thought Sandra. "All I know is I've done my best. I mean, that kid is my life. I just don't know what to do."

"You seem to be blaming yourself for Frankie's difficulties."

"Well, who the hell else is there to blame?" Sandra asked. "The school blames me, the therapists blame me, his caseworker blames me. I'm elected." Sandra was a little surprised at how angry she was.

"We don't blame people around here," said Ms. Brennan. "We assume parents do the best they can."

Sandra pondered this. "Well, my best clearly wasn't good enough."

"We see a lot of that around here," said Ms. Brennan.

"A lot of what?"

"A lot of parents who've tried really hard but don't have much to show for it."

"This lip is what I have to show for it," said Sandra.

"I think that kids like Frankie require a pretty specialized approach," said Ms. Brennan. "A lot of what seems to work for most kids doesn't work for kids like Frankie."

"Don't take this personally, but we've had a lot of people try to help us. And I'm still getting hit. So I'm kind of skeptical."

"I don't blame you. But I think one thing is obvious. We really need to help you and Frankie start talking to each other, but in a way that doesn't cause you to get hit."

"That would be good. I just . . . I don't know if it's very likely."

"What do you try to talk to him about?"

"School mostly," said Sandra. "I mean, we live in a pretty small apartment, so there's all the crap that goes along with living in a tight space. How loud he plays his music, how much time he spends playing video games, what video games he's playing, leaving his dirty clothes on the floor. But it's mostly school. I'm really worried that he's going to get thrown out of the program he's in, and I don't know what we're going to do if that happens."

"How do you try talking with him about those things?"

"Well, like I said, I mostly try not talking with him about those things," said Sandra, "because I don't want it to get ugly. But the more I think about it, the more worried I get, and then I can't help but talk to him about it, but by then I'm so stressed out that I'm probably not very calm when I do it, and then it goes downhill pretty fast."

"If I was to ask Frankie about why you and he don't talk anymore, what do you think he'd say?" asked Ms. Brennan.

"He'd say I don't listen," said Sandra. "He says that all the time. Maybe he's right."

"We're going to find out," said Ms. Brennan. "But I'd like to teach you a way to solve those problems with him—school, the loud music, the laundry, the video games—that I'm pretty sure won't cause him to feel that you're not listening to him. We're going to need lots of information from Frankie for those problems to get solved. Once they're solved—and once you and Frankie are able to solve problems together—I don't think you're going to get hit anymore."

Ms. Brennan described Plan B and its three steps. "So, what I'd like to do right now is try talking with Frankie about a problem, and I think the fact that he hates his program at school is a good place to start. Not that it's the only thing that we need to talk with him about, but it sounds like it's a major source of conflict for you two. All I'm going to do is the Empathy step. You can join in if you'd like, but the main thing is for you to see what drilling looks like. We really want to understand his concerns about that unsolved problem."

"OK," said Sandra.

Frankie shuffled in with a staff member. He meekly said hi to Sandra.

"Hi, Frankie," said Sandra. "Do you have everything you need here?"

Frankie nodded. "I'm sorry I hit you."

"I know."

"I just needed you to stop talking," said Frankie.

"I think," said Ms. Brennan, "that I might be able to help you and your mom talk together in a way that works better."

"I don't like talking to my mom."

"How come?" asked Ms. Brennan.

"She's . . . she gets too flipped out about everything."

"Like what?"

Frankie sighed. "She always stressed about money, and work . . . and me. It's just easier if we don't talk."

"From what I can gather, there are a lot of things that you should probably be talking about," said Ms. Brennan.

"Yeah, but not to her."

"To who?"

"I don't know," said Frankie. "Not her."

"So I heard you say that she's too stressed out about everything. How does that make it hard for you to talk to her?"

"She doesn't listen," said Frankie. "She just kinda barges in my room and goes totally crazy on me. It's not talking."

"Shall we see if your mom can listen now while I talk to you about school?" asked Ms. Brennan.

"I don't really see the point," said Frankie. He looked at Sandra. "You're the one who agreed for me to be in that freaking program. You didn't even ask me."

Sandra wasn't sure what to say. She looked to Ms. Brennan for help. "Go ahead," Ms. Brennan encouraged.

"Frankie, I agreed for you to be in that program because the people at school said it was the best thing for you. I guess it was the wrong decision. But I didn't know what else to do. I've never known what to do." Sandra started tearing up.

Frankie looked at Ms. Brennan. "I'm not doing this."

"How come?" Ms. Brennan asked.

"She's already crying. I don't wanna deal with that crap. That's why I like talking to the staff here. They listen, they don't freak out."

Sandra covered her face.

"See what I mean?" said Frankie, jumping out of his chair. "I don't want to do this!"

"You don't have to do it," said Ms. Brennan. "But hear me out for a second. Then you can decide if you want to stay."

Frankie stood by the door of Ms. Brennan's office.

"I could be wrong, but I have a feeling your mother is tougher than you think. Her life hasn't been easy."

"I know her life hasn't been easy! I thought we were talking about my life."

"Let's do that," said Ms. Brennan. "All I'm saying is that I think your mom can listen to what you have to say."

"Without crying? Or yelling?"

"I can't promise you that she won't cry. I don't think she's going to yell because you'll be talking to me. Your mom is just going to listen."

Frankie was silent.

"Can you talk to me about the things you don't like about your school?" asked Ms. Brennan.

"Yes."

"Can your mom stay in the room while we're talking?"

"If she keeps crying, I'm leaving," said Frankie.

"That's fine," said Ms. Brennan. She turned to Sandra. "Can you listen to what Frankie has to say about his school without crying?"

"I'll try," said Sandra.

Frankie sat back down.

"So, tell me about the difficulties you've been having at school. What's going on with that?"

"I want to go back to my regular junior high school," said Frankie, glancing warily at Sandra.

"That's good to know," said Ms. Brennan, recognizing that Frankie had voiced a solution rather than a concern. She steered the discussion back to Frankie's concerns. "I hear that's what you want to do, but I don't understand what it is that's making you want to do that."

"The kids in my program are freaks. And the teachers are losers. And I don't want to be a speddie anymore," said Frankie, using the slang for a child receiving special education services. "I wanna be normal."

"That's a lot of reasons," said Ms. Brennan.

"And I don't want my Mom making decisions about me without

me." Frankie glanced at Sandra again.

"OK," said Ms. Brennan. "Is there anything else you want to say about any of those things?"

"Not really," said Frankie.

"What do you mean that you don't want to be a speddie anymore?" asked Ms. Brennan.

"I'm sick of the whole thing. I've been getting extra help for writing and math since I was in freaking elementary school. I still suck at writing and math. If I'm still gonna suck at writing and math, I can do it at my old school and not be stuck with a bunch of freaks who are way more messed up than I am. And I don't need to go to a speddie school to get suspended. I was getting suspended at my old school too. At least at my old school I had friends."

"What's hard for you about writing and math?" asked Ms. Brennan.

Frankie looked down. "I don't know," he said softly. "I just . . . can't do . . . what other kids can do. I never could."

Sandra thought she saw a tear roll down Frankie's cheek. He quickly wiped it away. Sandra hadn't seen Frankie cry since elementary school. The sight made her eyes well up.

"Um, I'm going to excuse myself for a minute," she said, getting out of her seat.

"Do you want your Mom to go out?" asked Ms. Brennan.

Frankie looked up at Sandra with glassy eyes. "I knew you couldn't do it," he said. "But you don't have to go out."

Sandra sat back down.

"It's helpful for us to hear about these things," said Ms. Brennan softly. "I think it's good that we're getting to know what problems need to be solved."

Frankie shook his head slowly. "I don't think any of these

things can be fixed," he said, looking weary.

"Help me understand one more thing," said Ms. Brennan. "I understand that you have difficulty with math and writing. But I don't understand how those things cause you to get into trouble at school."

"Because teachers give me a hard time if I can't do assignments," said Frankie. "I'm not gonna sit there and get embarrassed in front of everyone. So I either say something rude to get them off my case or I just walk out of the class."

"Got it," said Ms. Brennan. She paused to think. "I think we have a large pile of problems that have been unsolved for a very long time. Different people have tried to help, but the problems are still there. And now I think you're both pretty pessimistic about whether they'll ever get solved. I'm very sorry those problems have made it so difficult for you to talk with each other and get along."

Frankie and Sandra were silent.

"I don't yet know what the solutions will be to the problems at school," said Ms. Brennan. "We'll have to figure that out together over the next few days, and whatever solutions we come up with here may need to be revisited if they aren't working. But here's what I do know. Frankie, I know that you're not signing off on any solution that doesn't address your concerns and doesn't have your input. That's happened a lot in your life, and it's made you very unhappy. It's also caused you and your mom to fight a lot and stop talking to each other. Sandra, I know that you have important concerns too, and we need to make sure that those concerns are heard and addressed as well. But what I mostly know is that the problems you and Frankie are facing don't have to cause conflict between you. Solving problems doesn't have to be adversarial. We just need to help you get good at solving problems together."

11

THE DINOSAUR
IN THE BUILDING

As hard as it is to help a behaviorally challenging child within a family, it may be even harder at school. After all, there are twenty or thirty other students in the child's classes, many with a wide range of special needs. Like parents, most general education teachers and school administrators have had no specialized training to prepare them to understand and help a behaviorally challenging child. Those who *have* received training probably learned more about Plan A than Plan B. There are a lot of different people working with the same child, and everyone's pressed for time. And there's a big dinosaur in the building: the school's discipline program.

Fortunately, many kids who are behaviorally challenging at home aren't challenging at school. This pattern often reinforces the false belief that a kid's challenging behaviors are intentional, goal-oriented, and completely under his control. Here are a few alternative explanations for the home/school disparity:

THE SITUATIONAL FACTOR: As you've read, challenging behaviors occur when the demands of the environment exceed a kid's capacity to respond adaptively. For some challenging kids, the demands of the school environment don't exceed their capacity to respond adaptively, but certain demands of the home environment do. For example, because the school environment tends to be relatively structured and predictable, it can actually be more "user-friendly" for some behaviorally challenging kids than the home environment. There are, however, behaviorally challenging kids who don't find the structure and predictability of school to be user-friendly at all; they're often the ones who have challenging episodes at school, too.

THE EMBARRASSMENT FACTOR: Many behaviorally challenging kids would be absolutely mortified if their classmates and teachers witnessed a challenging episode, so they put extraordinary energy into holding it together at school. Since the potential for embarrassment decreases at home, and since the energy can't be maintained 24/7, the kid unravels when he gets home. Most of us are better behaved outside the home than we are inside, so challenging kids aren't especially unusual in this regard. Of course there are challenging kids whose frustration at school blows right through the embarrassment factor.

THE CHEMICAL FACTOR: Teachers and classmates are often the primary beneficiaries of pharmacotherapy because the behaviorally challenging child is medicated during school hours. But the effects of many medications wear off by late afternoon or early evening, just in time for challenging episodes at home.

The fact that challenging episodes aren't occurring at school doesn't mean that unsolved problems at school aren't contributing to episodes that occur elsewhere. Many things at school can fuel episodes outside of school: being teased or bullied by other kids, social isolation or rejection, frustration and embarrassment over academic struggles, feeling misunderstood by a teacher. Homework often extends academic frustrations well beyond the end of the school day. So schools still have a role to play in helping behaviorally challenging kids, even if they don't see the kid at his worst.

This chapter focuses on kids who do have challenging episodes at school. Everything you've read in this book so far is applicable to schools and classrooms. But implementation at school isn't easy. Most school discipline programs are oriented toward Plan A; intervention for behaviorally challenging students often occurs in the heat of the moment rather than proactively; teacher and school evaluations are based primarily on the performance of their students on high-stakes, mandated testing with no emphasis on students' social, emotional, and behavioral well-being; budgets are extremely tight; and time is short. In many instances, teachers justifiably feel that they lack the expertise and are not provided with the support they need to understand and help kids with social, emotional, and behavioral challenges. What a shame, since challenging behavior deserves the same compassion and effort as any other developmental delay.

To make things still worse, misguided, ineffective zero-tolerance policies have driven many schools to use discipline rubric systems, which are usually comprised of a list (often a long one) of things students shouldn't do and an algorithm of adult-imposed consequences attached to each problem behavior on the list. But years of research is crystal clear on two points: zero tolerance policies have made things worse, not better; and standard school disciplinary practices generally aren't effective for the students to whom they are most frequently applied. The school discipline program isn't the reason well-behaved students behave well: they behave well *because they can*. We have little good to show for the millions of punishments—detentions, suspensions, paddlings, and expulsions—that are meted out every year to the kids having difficulty handling the social, emotional, and behavioral expectations at school. And yet most administrators' standard rationale for the continued use of consequences goes something like this:

Even if suspension doesn't help Frankie, at least it sets an example for our other students. We need to let them know that we take safety seriously at our school.

QUESTION: What message do we send the other students if we continue to apply interventions that aren't helping Frankie behave more adaptively?

ANSWER: That we're actually not sure how to help behaviorally challenging kids behave more adaptively.

QUESTION: What's the likelihood that the students who aren't behaviorally challenging would become behaviorally challenging if we did not make an example of Frankie?
ANSWER: Slim to none.

QUESTION: What message do we give Frankie if we continue to apply strategies that aren't working?
ANSWER: We don't understand you and we can't help you.

QUESTION: Under which circumstance do we have the best chance of helping Frankie solve the problems that are setting in motion his challenging episodes at school: when he's in school, or suspended from school?
ANSWER: When he's in school.

QUESTION: Why do many schools continue to use interventions that aren't working for their behaviorally challenging students?
ANSWER: Because they aren't sure what else to do and because popular contemporary disciplinary initiatives are not a dramatic departure from business-as-usual.

QUESTION: What happens to students to whom these interventions are counterproductively applied for many years?
ANSWER: They become more alienated and fall farther outside the social fabric of the school.

QUESTION: Isn't it the parents' job to make their child behave at school?

ANSWER: Helping a child deal more adaptively with frustration is everyone's job. The parents aren't there when the child has challenging episodes at school.

QUESTION: Isn't it the job of special education to handle these children?

ANSWER: Most behaviorally challenging kids don't need special education. Their developmental delay is in skill areas (flexibility, frustration tolerance, problem solving) that can usually be addressed in general education.

QUESTION: What usually happens to behaviorally challenging students when we apply a sink-or-swim mentality?

ANSWER: They sink.

Time for Plan B. But solving problems collaboratively in a school is no small undertaking. Here are some of the necessary components:

AWARENESS: Students with social, emotional, and behavioral challenges are being ill served by the disciplinary practices in many schools. Some educators know this already and are eager to learn new ways of understanding and helping these kids. Other educators need to be enlightened.

URGENCY: Understanding and helping these students has to be a priority. However, since educators have so many competing priorities, helping behaviorally challenging students often sits low on the totem pole. But we're losing a lot of kids unnecessarily because their behavioral challenges are misunderstood and mishandled.

MENTALITY: An adult's mentality or philosophy about children is what guides and governs his or her response when a student is not doing well. Many schools have adopted a *kids do well if they can* mentality. Regrettably, many are still stuck in the *kids do well if they want to* rut.

EXPERTISE: Many schools have been using the same discipline strategies for decades, blind to the fact that the "frequent fliers"—the students who are chronically in the office, or in detention, or being paddled or suspended—don't benefit from those strategies. They've yet to reckon with the fact that those strategies don't address the true difficulties (lagging skills and unsolved problems) of students with behavioral challenges. Some educators believe that the expertise necessary for understanding and helping behaviorally challenging students is well beyond their grasp. Not true. Educators need expertise and experience in two realms: identifying lagging skills and unsolved problems, and using Plan B. The expertise comes from reading books like this one. The experience comes from practice, which, first and foremost, requires effort and courage. Proficiency comes after that.

ABANDONING BLAME: It's time to stop blaming parents for challenging behavior that occurs at school. While it is true that some behaviorally challenging kids go home to family situations that are not ideal, it is also true that many *well*-behaved students come from family situations that are not ideal. Blaming parents is a counter-productive dead end, and it makes it much harder for school staff to focus on the things they can actually do something about: unsolved problems and lagging skills. Parents of behaviorally challenging kids get much more blame than they deserve for their kids' difficulties, just as parents of well-behaved kids get much more credit than they deserve for their kids' positive attributes.

TIME: Classroom teachers often feel that they don't have time to help kids with social, emotional, and behavioral challenges. But solving a problem collaboratively and proactively always takes less time than leaving the problem unsolved. Time is almost always a major concern *before* teachers and administrators learn how to use Plan B, but those concerns fade once educators become skilled at Plan B and embrace it. Indeed, the common mantra at schools where solving problems collaboratively is at the core of efforts to help behaviorally challenging students is as follows: *It takes time to do Plan B, but Plan B saves time.* When do staff members in these schools carve out time to use Plan B? Sometimes before school, sometimes after school, sometimes during lunch, sometimes during recess, sometimes during the teacher's prep time. I've yet to meet the administrator who isn't willing to arrange for coverage so that a classroom teacher can use Proactive Plan B with an individual student. Some schools have found it worthwhile to rethink the daily schedule to create the time needed for helping kids who would otherwise become lost in the shuffle.

ASSESSMENT MECHANISMS AND TOOLS: It's necessary to achieve consensus on the lagging skills and unsolved problems of each behaviorally challenging student so that the factors underlying his difficulties are well understood and the problems to be solved are clear. Identifying lagging skills and unsolved problems usually requires a meeting or two involving all of the adults who interact with the child at school, and the *Assessment of Lagging Skills and Unsolved Problems* should be the standard discussion guide in these meetings. Many schools have also begun incorporating lagging skills and unsolved problems into their functional assessments. It's crucial to go beyond merely concluding that a student's challenging behavior gets him something he wants (for example, attention), allows him to escape and avoid tasks and situations that are difficult, uncomfortable, tedious, or scary, and is therefore "working." A good functional assessment needs to explain why a student is going about getting, escaping, and avoiding in such a maladaptive fashion (lagging skills) and when that is occurring (unsolved problems).

PRACTICE, FEEDBACK, AND COACHING: Once mechanisms for assessing lagging skills and unsolved problems are in place, the next step is to become proficient at Plan B, a process that will require practice and ongoing feedback and coaching. The aspects of Plan B that are hard for parents tend to be challenging for educators as well: drilling for information in the Empathy step, identifying and articulating the adults' concern or perspective, and brainstorming solutions and considering whether each is realistic and mutually satisfactory. After numerous attempts, adults come to recognize their own unique vulnerabilities in using Plan B. For example, many stumble in their use of Plan B because of preconceived notions about kids' concerns, which can make it difficult to drill for information in an unbiased fashion. Other adults are inclined toward preordained solutions, which can make it hard to explore the range of mutually satisfactory and realistic possibilities.

ONGOING COMMUNICATION: Because Proactive Plan B is far preferable to Emergency Plan B, advance preparation and good communication among adults are essential. The only treatment models that don't require good communication are the *ineffective* ones. To help out, another instrument, the *Problem Solving Plan* can be found at www.livesinthebalance.org. It was designed to help adults keep track of high-priority unsolved problems for individual students, who is taking primary responsibility for solving each problem with the student, and the progression of problem-solving efforts through the steps of Plan B.

CONTINUITY: In schools, as in homes, there's a tendency to work on the hot-button problem that precipitated a challenging episode on a particular day. But because unsolved problems wax and wane, the hot-button unsolved problem that was the focal point on one day is often replaced by a different hot-button unsolved problem the next. The *Problem Solving Plan* is designed to prevent "problem-hopping" by helping adults track unsolved problems over time until they're durably resolved. The need for ongoing monitoring means that the adults working with a given child must reconvene periodically to assess progress and revisit unsolved problems.

PERSEVERANCE: Show me a behaviorally challenging student who people are trying to fix quickly, and I'll show you a behaviorally challenging kid it's taking a very long time to help. There is no quick fix. You're in this for the long haul. You don't fix a reading disability in a week, and you don't fix the developmental delays that contribute to challenging behavior in a week either. There will be bumps in the road. Transforming school discipline is a project. It doesn't happen overnight. But it needs to happen.

Naturally, there's much more that could be said about each of these components. That's why I wrote *Lost at School*, which was published in 2008.

Plan B isn't limited to adult–child problem solving. The ingredients of Plan B are equally applicable to unsolved problems between two kids and to those that affect an entire group of kids. And Plan B has significant ramifications for adult–adult problem solving as well. For the remainder of this chapter, let's consider what Plan B looks like when applied to these different types of problem solving in a school setting. We'll start with Plan B involving a teacher and student, move on to problem solving in a group of students, and finish with parents and teachers.

STUDENT–TEACHER PROBLEM SOLVING

As you'll see, Proactive Plan B doesn't look much different when the adult is a teacher rather than a parent. The ingredients are exactly the same though the topics may differ. Here's an example between a teacher and a thirteen-year-old student:

TEACHER: Class, please get to work on your social studies workseets.

RICKEY: I'm not doing it.

TEACHER: Well, then your grade will reflect both your attitude and your lack of effort.

RICKEY: I don't give a damn about my grades. I can't do this crap.

TEACHER: Your mouth just bought you a detention, young man. And I don't want students in my classroom who don't do their work. Anything else you'd like to say?

RICKEY: Yeah, this class sucks.

TEACHER: Nor do I need to listen to this. You need to go to the assistant principal's office now.

Oops. That was Plan A, wasn't it? Tricky author. Since this was an emergent problem, the teacher had much better options: Emergency Plan C or Emergency Plan B. Here's what Emergency Plan C would look like:

TEACHER: Class, please get to work on your social studies worksheets.

RICKEY: I'm not doing it.

TEACHER: You're not doing it.

RICKEY: Forget it. I can't do this! Just leave me alone! Damn!

TEACHER: Rickey, you don't have to do it. But hang on for just a second. Let me get everyone else going, and then you and I can figure out what's the matter and see what we can do about it.

And here's what the same problem would look like if it were handled with Emergency Plan B:

TEACHER: Class, please get to work on your social studies worksheets.

RICKEY: I'm not doing it.

TEACHER: Tell me what's going on, bud.

RICKEY: Forget it. I can't do this! Just leave me alone! Damn!

TEACHER: Rickey, tell me what's going on.

RICKEY: You know I have trouble with the spelling!

TEACHER: Yes, I do know that. That's why I don't grade you for the spelling.

RICKEY: But it still bugs me.

TEACHER: I get it. The thing is, I need a way to know what you've learned about George Washington Carver.

I wonder if there's a way for you to let me know what you've learned about George Washington Carver without you getting frustrated about the spelling part.

RICKEY: How?

TEACHER: I don't know, let's think about it.

RICKEY: I don't have any ideas!

TEACHER: Well, maybe Darren would help you with any words you don't know how to spell on the worksheet.

RICKEY: No way.

TEACHER: How come?

RICKEY: He's going to rag on me about needing his help.

TEACHER: Hmm. Is there anyone who could help you who wouldn't give you a hard time about it?

RICKEY: DeJuan.

TEACHER: DeJuan? That could work. You'd feel more comfortable with him?

RICKEY: Yeah, he's smart.

TEACHER: You're smart, too. You just have some trouble with spelling.

While Emergency Plan C and Emergency Plan B are useful for defusing eruptions, Rickey's spelling problem is predictable. Rather than dealing with his spelling problem emergently, every day, which is time-consuming, the teacher would want to schedule a time to work on the problem with Rickey using Proactive Plan B, preferably before it erupts in the middle of another lesson.

By the way, the ingredients of Plan B can be applied to every student in the class, each of whom has problems that need to be solved. If every student is working on something, then the behaviorally challenging kid isn't singled out—you're solving problems collaboratively with everyone.

STUDENT–STUDENT AND GROUP PROBLEM SOLVING

Plan B can be also applied to unsolved problems that arise between two students. In such instances, the teacher's role is Plan B facilitator. Here's an example from *Lost at School*:

MR. BARTLETT: Hank, as you know, in our classroom, when something is bothering somebody, we try to talk about it. As I mentioned to you yesterday, I thought it might be a good idea for me and you and Laura to talk about the project you guys are supposed to be doing together.

HANK: OK.

MR. BARTLETT: She has some concerns about what it's going to be like doing the project with you. It sounds like you guys worked on a project together last year, yes?

HANK: Yup.

MR. BARTLETT: I don't know if you knew this, but Laura came away from that project feeling like you weren't very receptive to her ideas and like she did most of the work. So she wasn't too sure she wanted to do this project with you.

HANK: She doesn't have to do the project with me. I can find another partner.

MR. BARTLETT: Yes, she was thinking the same thing. But I was hoping we could find a way for you guys to work well together. What do you think of Laura's concern?

HANK: I don't know. That was a long time ago.

MR. BARTLETT: Do you remember how you guys figured out what to do on last year's project?

HANK: No.

MR. BARTLETT: Do you remember that Laura did most of the work?

HANK: Sort of. But that's because she didn't like the way I was doing it, so she decided to do it herself.

LAURA: That is so not true. I did most of the work because you wouldn't do anything.

HANK: Well, that's not how I remember it.

MR. BARTLETT: It sounds like you both have different recollections about what happened last year and why it didn't go so well, so maybe we shouldn't concentrate on what happened last year. I don't know if you will ever agree on that. Maybe we should focus on the concerns that are getting in the way of your working together this year. Laura, your concern is that Hank won't listen to any of your ideas. And you're both concerned about the possibility that Laura will do all the work. I wonder if there's a way for you guys to make sure that you have equal input into the design of the project, without having Laura do all the work in the end. Do you guys have any ideas?

LAURA: This is so pointless. He won't listen to my ideas.

MR. BARTLETT: Well, I know that's what you feel happened last year, but I can't do anything about last year. We're trying to focus on this year and on coming up with a solution so that you and Hank have equal input and work equally hard.

LAURA: Can you sit with us while we're figuring out what to do? Then you'll see what I mean.

HANK: Then you'll see what I mean.

MR. BARTLETT: So, Laura, you're saying that maybe if I sit in on your discussions I might be able to help you guys have a more equal exchange of ideas?

LAURA: That's not really what I meant.

MR. BARTLETT: I know, but I'm thinking that it might not be a bad way to ensure the equal exchange of ideas. What do you think?

HANK: I think we can work together.

LAURA: Fine, sit in on our discussion and help us have equal input.

MR. BARTLETT: Only if that works for you guys.

LAURA: It only works for me if I have to work with him.

MR. BARTLETT: I'm not saying you have to work with him. I'm saying I'd like you to give it a shot so the other kids don't have to break up their pairs. We can entertain other options if that solution doesn't work for you.

LAURA: What other solutions?

MR. BARTLETT: I don't know. Whatever we come up with. Can you guys think of any others?

HANK: We could do the project by ourselves, you know, alone. She could do one, and I could do one.

MR. BARTLETT: Well, that would probably work for you guys, but it wouldn't work for me. One of the goals of this project was for kids to learn to work together. I think it's an important skill.

LAURA: Why don't we try to work together, with you helping us, and if that doesn't work we can do our own projects.

MR. BARTLETT: Hank, does that solution work for you?

HANK: Sure, whatever.

MR. BARTLETT: I need to think about whether it works for me. You guys will try hard to work together with me helping you?

LAURA: Yes.

HANK: Yes.

MR. BARTLETT: OK, let's go with it. We're working on the project again tomorrow. I'll sit in on your discussion with each other and see if I can help make sure the exchange of ideas is equal and the workload is equal. Let's see how it goes.

While some problems are best addressed by using Plan B with individual students or pairs of students, other problems, especially those that affect the group as a whole, are best addressed by involving the entire class in a Plan B discussion. Students are usually already accustomed to group discussions on academic topics. When Plan B is added to a group discussion, and when such discussions are about nonacademic problems like bullying, teasing, and classroom conduct, then community members learn to listen to and take into account one another's concerns. Group problem solving is difficult, but no harder than having problems that remain unsolved.

Again, the ingredients are the same, and the teacher is the facilitator. The first goal is to achieve the clearest possible understanding of the concerns and perspectives of students regarding a given problem. Once their concerns have been clarified, the group moves on to finding a solution that will address those concerns. The criteria for a good solution remain the same: it must be realistic and mutually satisfactory.

When using Plan B with a group, the teacher helps the group decide what problems to tackle first, keeps the group focused and serious, and ensures that the exploration of concerns and solutions is exhaustive. (Eventually the students will take on these responsibilities themselves.) There are no good or bad concerns, no such thing as "competing" concerns, only concerns that need to be addressed. Likewise, there are no right or wrong solutions, only solutions that are realistic (or not) and mutually satisfactory (or not).

PARENT–TEACHER PROBLEM SOLVING

Parents of behaviorally challenging kids and school personnel often have difficulty working together for the same reasons that kids and adults do: a tendency to blame one another; a failure to achieve a consensus on the true nature of a kid's difficulties (lagging skills) and the true events (unsolved problems) that precipitate his challenging episodes; a failure to identify the concerns of the respective parties; and the attempt of one party to impose its will on another. As Sarah Lawrence-Lightfoot writes in her insightful book *The Essential Conversation: What Parents and Teachers Can Learn from Each Other*, great potential exists for productive collaboration between parents and teachers. When parents and teachers are able to exchange specific information about a child's lagging skills and unsolved problems, they start trusting each other. Parents become convinced they are being heard and that the teacher sees, knows, and cares about their child. Educators become convinced that the parents are eager to receive information, collaborate, and help in any way possible. Both parties need to be involved in the process of working toward a mutually satisfactory action plan. You're on the same team.

Here's what Proactive Plan B looks like between parents and teachers. Once again, it uses the same familiar ingredients: information gathering and understanding, considering both parties' concerns, and brainstorming solutions that are realistic and mutually satisfactory.

TEACHER: I understand that homework has been very difficult lately.

MOTHER: Homework has been very difficult for a very long time. You're the first teacher Rickey's had who's expressed any interest in what we go through with homework. We spend several hours struggling over homework every weeknight and every weekend.

TEACHER: I'm sorry about that. But let's see if we can figure out what's so hard about homework and then come up with a plan so it's not so terrible anymore.

MOTHER: You can't imagine how nice that would be.

TEACHER: Can you tell me the parts of homework that have been difficult for you and Rickey? Or is it all hard? You don't mind if I write these down, do you?

MOTHER: Not at all. He's a very slow writer. So he gets frustrated that homework takes as long as it does.

And he seems to have trouble thinking of a lot of the details you're asking for. And he's always struggled with spelling. Last year's teacher told us not to worry about the spelling. But Rickey doesn't seem to be able to do that. So I don't know whether to forget about it or work on it. I wouldn't know how to work on it anyway. So I end up doing a lot of the writing for him.

TEACHER: Yes, I've noticed that he writes slowly, and the difficulty he has coming up with details, and his troubles with spelling. How about math?

MOTHER: He breezes right through it. Very little writing, very little spelling, and not the kind of details he has trouble with.

TEACHER: Well, then let's take our problems one at a time. Of course I've only had Rickey in my class for about four weeks now, so I can't say I have a perfect handle on his difficulties or what we should do about them. And I've begun working with Rickey on these problems myself, so I'm in the midst of trying to gather some information from him, too. But I'm not one for having kids spend two hours on homework every night, and certainly don't want homework to cause problems between kids and their parents. Of course, I'm not always aware that those problems exist, so I appreciate your honesty.

MOTHER: I'm not shy about letting people know what's going on with Rickey. I just wish we were seeing more progress on the problems he's having.

TEACHER: The thing is, we're going to need to get Rickey involved in the homework discussion, too. Even if you and I come up with brilliant solutions, they won't be so brilliant if he's not on board with them. So maybe we should use this discussion to make sure we have a clear sense of the problems we need to solve. One problem is the amount of time homework is consuming. Yes?

MOTHER: Yes.

TEACHER: But it sounds like a lot of that time is spent being frustrated over what to work on and how you can help, so that's something we'll need to solve, too.

MOTHER: Absolutely.

TEACHER: I'm not convinced that Rickey can't get better at spelling, so I'm disinclined to tell you that we should drop it altogether. Plus, as you said, Rickey doesn't seem able to drop it. So spelling is an unsolved problem. And slow writing is an unsolved problem. And fleshing out the details is an unsolved problem. And I know you're doing a lot of the writing for him, but we don't want him getting the idea that he doesn't need to do any of the writing.

MOTHER: Aren't you overwhelmed by all this?

TEACHER: No, I actually find that sorting through unsolved problems helps me become less overwhelmed. At least I know what needs to be addressed.

MOTHER: I see what you mean.

TEACHER: Any other unsolved problems related to homework?

MOTHER: Well, he has football practice two nights a week, so sometimes he's really tired when it's time for homework. Those are our really tough nights.

TEACHER: I can imagine. So we have some work to do, don't we?

MOTHER: It appears so.

TEACHER: Here's what I'm thinking. If it's OK with you, why don't we meet again next week, but next time let's include Rickey in the meeting. Then we can start talking about how these problems can be solved, one at a time.

What's the solution to the writing problem? The spelling problem? The details problem? The football practice problem? That's for Rickey, his mom, and his teacher to figure out. There are dozens of possibilities. There's no such thing as a right or wrong solution—only solutions that are realistic and mutually satisfactory. What will they do if the first solution to a given problem doesn't stand the test of time? They'll head back to Plan B, figure out what didn't work—in other words, what it was about the solution that wasn't realistic and mutually satisfactory—and come up with a better solution.

Q & A

QUESTION: I'm a teacher, and I'm a little worried about having different expectations for different kids. If I let one kid get away with something, won't my other students try to get away with it as well?

ANSWER: Plan B isn't about letting students get away with something. Teachers *usually* have different expectations for different children. That's why some students receive special reading help while others do not and why some students are in a gifted program for math while others are not. If a student asks why one of his classmates is being treated differently, it gives the classroom teacher the perfect opportunity to do some educating: *"Everyone in our classroom gets what he or she needs. If someone needs help with something, we all try to help. And everyone in our class needs something special."* It's no different when a child needs help with flexibility, frustration tolerance, and problem solving.

QUESTION: Does Plan B undermine a teacher's authority with the other kids in the class?

ANSWER: No, it doesn't. The other kids are watching closely. If a teacher intervenes in a way that solves problems, teaches skills, and reduces the likelihood of an explosion, he or she has done nothing to undermine his or her authority with the other kids.

QUESTION: Is it really fair to expect teachers—who are not trained as mental health professionals—to solve all these problems with their students?

ANSWER: There is much that is unfair about the whole picture. It's not fair that school staff are pressured to prioritize the academic well-being of their students over kids' social and emotional well-being; those priorities should go hand-in-hand. It's not fair that because those priorities are misplaced, teachers don't receive the training and support they need to understand and help their behaviorally challenging students and must still rely on obsolete, ineffective discipline strategies. It's not fair that other students are adversely impacted by students whose behavioral challenges have been poorly understood and poorly handled for a long time. And it's not fair that an astounding number of behaviorally challenging kids are still needlessly slipping through the cracks.

A mental health degree is not a prerequisite for solving problems collaboratively: most mental health professionals don't have training in solving problems collaboratively, either. The key qualifications for helping kids with behavioral challenges are an open mind, a willingness to reflect on one's current practices and see them in a new light, the courage to try new practices, and the patience and resolve to become comfortable assessing lagging skills and unsolved problems and using Plan B.

QUESTION: Are there school personnel who refuse to participate in learning about solving problems collaboratively because it goes beyond what they are paid to do?

ANSWER: Yes, but it's much more common for school personnel to be willing to go the extra mile to learn about new ways to help kids.

QUESTION: I was using Plan B with a kid in my class and things seemed to be going well for a few weeks but then deteriorated again. What happened?

ANSWER: It could be that the solution you and the kid agreed on wasn't as realistic and mutually satisfactory as it originally seemed. That's not your signal to revert back to Plan A; it's your signal to go back to Plan B to figure out why the solution didn't work as anticipated and to collaborate on a revised solution.

QUESTION: Are there some challenging kids who are so volatile and unstable that academics should be deemphasized until things are calmer?

ANSWER: Yes. Some kids simply aren't available for academic learning until they've made headway on the challenges impeding learning. Soldiering on with academics when a kid is bogged down in behavioral challenges is usually an exercise in futility.

QUESTION: What if Plan B isn't working? What then?

ANSWER: The answer depends on your definition of "working." For many people, "working" refers only to the ultimate destination, the point at which a problem is finally durably solved. But there are many ways in which Plan B is working before the ultimate destination is reached. Plan B is working if adults are viewing a kid's difficulties more accurately and more compassionately. It's working if adults are effectively gathering information about a kid's concerns on a given problem and finally understanding what's been getting in the kid's way. It's working if the kid is able to listen to adult concerns and take them into account. Plan B is working if the kid no longer views adults as the enemy. It's working if the kid is participating in discussions about how a given problem can be solved in a way that addresses the concerns of both parties. Plan B is even working if the kid isn't really participating yet fully but perceives that adults have shifted away from being unilateral and punitive and are trying really hard to understand and help.

QUESTION: Are there some students who need more than what can be provided in a general education setting, even if people are using Plan B?

ANSWER: Yes, there are. But wouldn't it be interesting to find out how many students still need more than what can be provided in general education settings if more general education settings used Plan B? That question aside, there are some kids who need a larger dose of Plan B than schools and outpatient settings can provide, kids who continue to behave in an unsafe manner at home, at school, and/or in the community. Many start a downward spiral early, become increasingly alienated, exhibit increasingly serious forms of inappropriate behavior, and begin to hang out with other kids who have come down a similar path. After all else has been tried—therapy, medication, perhaps even alternative day-school placements— these kids ultimately need a change of environment. A new start. A place to start forging a new identity. Once alienation and deviance become a kid's identity, things are a lot harder to turn around. Fortunately, there are some outstanding therapeutic day schools and residential facilities that do an exceptional job of working with such kids.

12

BETTER

You've made it to the last chapter, and you've covered a lot of ground along the way. The first goal was to help you view your behaviorally challenging kid more accurately, beginning with new thinking: *Kids do well if they can.* You now know that your kid's challenging episodes occur when the demands being placed on him exceed his capacity to respond well, and that if he could respond well, he would. We shelved a lot of conventional wisdom about challenging episodes (that they're intentional, goal-oriented, and purposeful), about behaviorally challenging kids (that they're unmotivated, attention-seeking, manipulative, coercive, and button-pushing), and about the parents of behaviorally challenging kids (that they're passive, permissive, and inconsistent disciplinarians). We examined the various lagging skills and unsolved problems that can set the stage for challenging episodes. You learned why traditional discipline, with its heavy emphasis on rewarding and punishing, hasn't improved your situation, and learned how to solve problems collaboratively (Plan B) and proactively rather than unilaterally (Plan A) and emergently. And you read about the different ways in which Plan B can go awry and how to get back on course. We've come a long way.

But I hope you went even further and began acting on all that new knowledge. Your first task was to identify the lagging skills and unsolved problems that apply to your situation. Your next task was to begin using Plan B to solve the problems that precipitate challenging episodes in your household.

My hope is that things are now better in your household. Things might have improved because you understand your child's difficulties better than you did before. Things might be better because you've removed some low-priority demands and expectations from the equation (Plan C). Maybe things are better because you're relying far less on Plan A and adult-imposed consequences to solve problems with your child. And things might be better because you and your problem-solving partner have collaboratively and proactively solved—one by one—a bunch of the problems that were causing challenging episodes. Hopefully, you're feeling like you and your child are communicating again and that your relationship is moving in the right direction.

Sometimes it's hard to notice that things are getting better. Some adults have a preconceived notion of what life is going to be like when things are finally "better" and are disappointed to find that life with their behaviorally challenging kid still isn't a walk in the park. Some wish it weren't so hard to improve things or that it could be accomplished more quickly. How quickly progress is made and how difficult it is differs for every behaviorally challenging kid and every family. And the definition of "better" is different for every behaviorally challenging kid and every family, too. For what it's worth, here's my definition of better: *it's better.* And better begets better.

Could things be even better than they are now? You'll find out. If you need more help, you know where to go: www.livesinthebalance.org.

And if you're the type of person who likes to read the entire book before putting what you've read into action, your time has come.

If you've been thinking, *"Shouldn't all children be raised this way?"* the answer is yes. While the model described in this book has its roots in the treatment of behaviorally challenging kids, clearly behaviorally challenging kids aren't the only ones who benefit from having their concerns identified and validated, taking another person's concerns into account, participating in the process of generating and considering alternative solutions to problems, working toward mutually satisfactory solutions, and resolving disputes and disagreements without conflict. *All* kids benefit . . . and so do all adults.

Sandra and Debbie were on the phone again. It had been a week since Frankie came home from the inpatient unit.

"How's Frankie doing?" asked Debbie.

"Well, he's going to be in the partial hospitalization program in Amberville for a few more weeks," said Sandra. "They're trying to help me get things sorted out with the folks at school to see if there's some way to get Frankie out of that program he hates. I don't know how that's going to pan out, but I feel like people are listening to me and helping me instead of just telling me what to do. One good thing—the door to his room isn't closed all the time. And he participated in a Plan B conversation yesterday—about playing his music too loud—right here in our apartment, with Matt helping out."

"Wow, that's progress," said Debbie. "You must be a little relieved."

"Oh, we have a long way to go," said Sandra. "*I* have a long way to go. I didn't know how to talk to my own kid. I didn't know how to solve his problems. I was trying, but I didn't know how. I was getting so worked up—especially about how things were going at school—that I let it get in the way. I was leaving out the most important person: Frankie. I'm seeing him in a very different way now. I think that may be the most important part. I feel like I'm starting to get my son back."

"I feel the same way about Jennifer," said Debbie.

"I haven't heard about Jennifer in the past few weeks," said Sandra. "I'm sorry."

"I think you've been a little busy," said Debbie. "Jennifer's talking. More with me than with Kevin, though he's trying hard. It's not easy for him to resist the temptation to rush through the Empathy step and head straight for solutions. But she corrects him when he does it."

"*She* tells *him* how to do Plan B?"

"Yeah, it's actually kind of funny. But we're finding out what's going on in that head of hers. She sure is rigid about some things. But we're getting some problems solved."

"That's wonderful."

"And she's letting me tuck her into bed at night again. She even let me hug her the other day without getting all pissed off about it."

"No! She did?"

"Well, I gave her advance warning that a hug was coming. She also screamed at me a few minutes later because I'd rearranged something in her room. So we have a lot more problems to solve." Debbie paused. "Do you think our lives will ever be normal?"

Sandra laughed. "My life hasn't been normal since the day I came into this world. I stopped shooting for normal a long time ago. The abnormal is my normal."

Debbie pondered this. "So normal isn't even the goal."

"I don't know what normal is," said Sandra. "I'm just concentrating on doing what I can to make tomorrow slightly better than today for me and my kid. That's what I've always done. I don't know what's around the bend. But I'm starting to feel like I know how I'll handle it—whatever it is—when I get there."

"And it's a lot easier when your kid is your partner instead of your enemy," said Debbie.

"Easier, for sure. Easy, no way."

I receive a lot of e-mail from parents and other care-givers. Many ask for help and guidance, others are seeking resources, and quite a few just want to let me know

258 The Explosive Child

how things are going with their child. The following e-mail, which I received from a father about five years ago, was especially memorable:

> This evening, after my twelve-year-old daughter stayed up late to finish a project for school, I couldn't help but reflect on how much she has changed in the past twenty months. Today she is a well-balanced student athlete with a great circle of friends. She demonstrates patience and good communication skills. Twenty months ago, she was certainly a behaviorally challenging child. We were quite certain that the only path to resolution was inpatient treatment. While we made some small advances in our understanding of her issues with local psychologists, we made few if any steps toward improvement. Then we read *The Explosive Child*. The issues and solutions became understandable and actionable. Without any professional help we implemented the solutions in the book, and over time the results have been amazing. I am writing to express my gratitude for providing the insights to restore normalcy in our lives. My child is back on the path to a productive and successful life. I have also learned a great deal about myself and human interaction in the process. I consider this the greatest accomplishment of my life.

Kids do well if they can. So do parents. And if things aren't going well for you and your behaviorally challenging child, now you know what to do.

ASSESSMENT OF LAGGING SKILLS & UNSOLVED PROBLEMS (Rev. 11-12-12)

Child's Name: _____   Date: _____

Instructions: The ALSUP is intended for use as a *discussion guide* rather than as a freestanding check-list or rating scale. It should be used to identify specific lagging skills and unsolved problems that pertain to a particular child or adolescent. If a lagging skill applies, check it off and then (before moving on to the next lagging skill) identify the specific expectations the child is having difficulty meeting in association with that lagging skill (unsolved problems). A non-exhaustive list of sample unsolved problems is shown at the bottom of the page.

LAGGING SKILLS	UNSOLVED PROBLEMS
__ Difficulty handling transitions, shifting from one mindset or task to another	
__ Difficulty doing things in a logical sequence or prescribed order	
__ Difficulty persisting on challenging or tedious tasks	
__ Poor sense of time	
__ Difficulty maintaining focus	
__ Difficulty considering the likely outcomes or consequences of actions (impulsive)	
__ Difficulty considering a range of solutions to a problem	
__ Difficulty expressing concerns, needs, or thoughts in words	
__ Difficulty understanding what is being said	
__ Difficulty managing emotional response to frustration so as to think rationally	
__ Chronic irritability and/or anxiety significantly impede capacity for problem-solving or heighten frustration	
__ Difficulty seeing the "grays"/concrete, literal, black-and-white, thinking	
__ Difficulty deviating from rules, routine	
__ Difficulty handling unpredictability, ambiguity, uncertainty, novelty	
__ Difficulty shifting from original idea, plan, or solution	
__ Difficulty taking into account situational factors that would suggest the need to adjust a plan of action	
__ Inflexible, inaccurate interpretations/cognitive distortions or biases (e.g., "Everyone's out to get me," "Nobody likes me," "You always blame me, "It's not fair," "I'm stupid")	
__ Difficulty attending to or accurately interpreting social cues/poor perception of social nuances	
__ Difficulty starting conversations, entering groups, connecting with people/lacking other basic social skills	
__ Difficulty seeking attention in appropriate ways	
__ Difficulty appreciating how his/her behavior is affecting other people	
__ Difficulty empathizing with others, appreciating another person's perspective or point of view	
__ Difficulty appreciating how s/he is coming across or being perceived by others	
__ Sensory/motor difficulties	

UNSOLVED PROBLEMS GUIDE: Unsolved problems are the specific expectations a child is having difficulty meeting. Unsolved problems should be free of maladaptive behavior; free of adult theories and explanations; "split" (not "clumped"); and specific.

HOME: Difficulty getting out of bed in the morning in time to get to school on time; Difficulty getting started on or completing homework (specify assignment); Difficulty ending the video game to get ready for bed a night; Difficulty coming indoors for dinner when playing outside; Difficulty agreeing with brother about what television show to watch after school; Difficulty handling the feelings of seams in socks; Difficulty brushing teeth before bedtime; Difficulty staying out of older sister's bedroom; Difficulty keeping bedroom clean; Difficulty clearing the table after dinner

SCHOOL: Difficulty moving from choice time to math; Difficulty sitting next to Kyle during circle time; Difficulty raising hand during social studies discussions; Difficulty getting started on project on tectonic plates in geography; Difficulty standing in line for lunch; Difficulty getting along with Eduardo on the school bus; Difficulty when losing in basketball at recess

©Ross W. Greene, Ph.D., 2012

ADDITIONAL READING

Helping Behaviorally Challenging Kids at Home

Parent Effectiveness Training: The Proven Program for Raising Responsible Children, by Thomas Gordon. Three Rivers Press, 2000.

Raising Your Spirited Child: A Guide for Parents Whose Child is More Intense, Sensitive, Perceptive, Persistent, and Energetic, by Mary Sheedy Kurcinka. Harper Paperbacks, 2006.

The Difficult Child, by Stanley Turecki and Leslie Tonner. Bantam, 2000.

Tired of Yelling: Teaching Our Children to Resolve Conflict, by Lyndon D. Waugh and Letitia Sweitzer. Pocket Books, 2000.

Helping Behaviorally Challenging Kids at School

Lost at School: Why Our Kids with Behavioral Challenges are Falling Through the Cracks and How We Can Help Them, by Ross W. Greene. Scribner, 2009.

Beyond Discipline: From Compliance to Community, by

Alfie Kohn. Association for Supervision and Curriculum Development, 2006.

Teaching Young Children in Violent Times: Building a Peaceable Classroom, by Diane E. Levin. Educators for Social Responsibility, 2003.

The Self-Control Classroom: Understanding and Managing the Disruptive Behavior of All Students Including Students with ADHD, by James Levin and John Shanken-Kaye. Kendall/Hunt Publishing, 2001.

The Challenge to Care in Schools: An Alternative Approach to Education, by Nell Noddings. Teachers College Press, 2005.

Parent–Teacher Interactions

The Essential Conversation: What Parents and Teachers Can Learn From Each Other, by Sarah Lawrence-Lightfoot. Random House, 2003.

Social Skills

Helping the Child Who Doesn't Fit In, by Stephen Nowicki and Marshall Duke. Peachtree Publishers, 1992.

Teaching Your Child the Language of Social Success, by Stephen Nowicki and Marshall Duke. Peachtree Publishers, 1996.

You Are a Social Detective: Explaining Social Thinking to Kids, by Michelle Garcia-Winner and Pam Crook. North River Press Publishing, 2010.

The New Social Story Book, Revised and Expanded 10th Anniversary Edition: Over 150 Social Stories That Teach Everyday Social Skills to Children with Autism or Asperger's Syndrome, and Their Peers, by Carol Gray. Future Horizons, 2010.

Sibling Interactions

Siblings Without Rivalry: How to Help Your Children Live Together So You Can Live Too, by Adele Faber and Elaine Mazlish. HarperCollins, 2012.

INDEX

ALSUP *(cont.)*
  and theories about causes of
    unsolved problems, 46
attention, parental, 192
  providing, 76
  withholding, 77
attention deficit hyperactivity
    disorder (ADHD), 13
attention seeking, as
    explanation for behavior,
    9, 35–36, 76, 90, 105,
    138, 253
  school and, 233
attitude, bad, 37
atypical antipsychotics, 172
autism spectrum disorders,
    13, 60–63

bad attitude, 37
bad choices, 37
battle picking, 96–97
bed, getting to, 22, 40, 103,
    109, 111, 118
behavior modification,
    conventional approach to,
    76–77
bipolar disorder, 2, 13, 63
black-and-white thinking,
    30–34
  sarcasm and, 205–6
blaming, 243
  of parents for problems at
    school, 232
bosses, 161

brushing teeth, 88–89, 90, 93,
    94, 96, 103, 109, 110, 117
button pushing, 9, 37, 75, 253

caregivers:
  tension between, 212–13
  see also parents
catastrophizing, 206
challenging behavior:
  beliefs about, 5, 12
  children's lack of choice in
    exhibiting, 11
  conventional wisdom about,
    13, 75–77, 82–83, 253
  focusing on skills vs., 11–12
  home/school disparity in,
    225–27
  information provided by,
    165
  lagging skills and, *see* lagging
    skills
  leaving out of unsolved
    problems, 44–45, 48, 52,
    81
  medications and, *see*
    medications
  Plan A and, 89–90, 98, 161,
    165, 166, 172
  predictable, 40, 42
  time needed to deal with,
    166
  as unpredictable, xii, 12, 40
  unsolved problems and, *see*
    unsolved problems

ABOUT THE AUTHOR

Photo by Elizabeth Humphreys Thomas

ROSS W. GREENE, PH.D., is Adjunct Associate Professor in the Department of Psychology at Virginia Tech, Associate Clinical Professor at Harvard Medical School, and the originator of the cutting-edge, empirically supported approach for understanding and helping kids with social, emotional, and behavioral challenges described in this book. He consults extensively to families, general and special education schools, inpatient and residential facilities, and systems of juvenile detention, and lectures widely throughout the world. Dr. Greene is the founding director of the nonprofit Lives in the Balance (www.livesinthebalance.org), which provides free, web-based resources on his model and advocates on behalf of behaviorally challenging kids and their caregivers. Dr. Greene lives in Portland, Maine, with his wife and two children.